Advances in Water and Wastewater Treatment
BIOLOGICAL NUTRIENT REMOVAL

Engineering and Industrial Experiment Station

Advances in Water and Wastewater Treatment
BIOLOGICAL NUTRIENT REMOVAL

edited by

Martin P. Wanielista, P.E.
Gordon J. Barnett Professor of
Environmental Systems Management at
Florida Technological University

W. Wesley Eckenfelder, Jr.
Distinguished Professor at
Vanderbilt University

From papers presented in part at a conference on
Biological Nutrient Removal Alternatives held
in Orlando, Florida at Florida Technological
University in March 1978.

230 Collingwood, P.O. Box 1425, Ann Arbor, Michigan 48106

Library of Congress Catalog Card No. 78-67495
ISBN 0-250-40282-3

Manufactured in the United States of America

PREFACE

Biological nutrient removal systems are relatively new technologies with the potential production of high quality effluents. In some areas, high quality effluents are required before discharge to receiving waters or before land application. Social and health problems associated with eutropic conditions in our nation's waters require greater removals of nutrients. These eutropic conditions have precipitated increased water pollution control requirements from point and nonpoint sources. At this time, in most cases, biological nutrient removal provides a means that can be most cost-effective. This reference volume has been planned to present state-of-the-art biological nutrient removal alternatives. This is done by incorporating results of recent research with design and operational details.

Eight of the chapters in this text were adopted from presentations at a conference on Biological Nutrient Removal Alternatives held at Florida Technological University in March 1978. Two papers are essentially the same as the presentations. Three chapters were added after the completion of the conference. The editors and authors express their gratitude for the support work received from their associates in the final compilation of the papers. The resources and services of the Florida Technological University are appreciated. In addition, the organizational help of the Florida Engineering Society and the Florida Section of the American Society of Civil Engineers was valuable in the conduct of the conference and the encouragement to publish this volume.

M.P.W. & W.W.E.

Martin P. Wanielista is a Professor of Engineering and the Gordon J. Barnett Professor of Environmental Systems Management at Florida Technological University. He is a graduate of the University of Detroit, Manhattan College and Cornell University, has conducted research in the area of stormwater management, and is a Registered Professional Engineer in the State of Florida.

Dr. Wanielista is the author of over 50 articles and technical reports and has edited or authored six books, including *Stormwater Management—Quantity and Quality* published by Ann Arbor Science. He has received 16 awards in six years for service in education, research and community leadership and is an active member of six professional and technical organizations.

Distinguished Professor of Environmental and Water Resources Engineering at Vanderbilt University, W. Wesley Eckenfelder, Jr. is a consultant to the UN and to Israel. He is also consultant to many industries, cities, countries and consulting firms. He is the author or editor of 10 books and over 200 technical papers in the field of water pollution control.

Professor Eckenfelder holds BCE, MS and MCE degrees and has received numerous honors. He is a member of the WPCF, ACS, TAPPI, ASCE, AIChE and other professional organizations and is listed in Who's Who in the South and Southwest and Who's Who in Engineering. He is a Fellow of the Institution of Public Health Engineers, an Honor Member of the International Association on Water Pollution Research, and a Fellow of the American Institute of Chemists.

TABLE OF CONTENTS

1. PROCESS SELECTION FUNDAMENTALS

W. M. McLellon, Civil Engineering and Environmental Sciences Department, Florida Technological University, Orlando, Florida

ABSTRACT

Selection of a process for wastewater treatment requires detailed evaluation of quality parameters of influent and effluent. The objective is to select a process train which will produce quality changes necessary, reliably, while under varying conditions of flow, and at minimum cost.

INTRODUCTION

In approaching a topic of this type, it is of interest in the beginning to consider the overall task. The purpose is to examine nutrient removal, which was, in the past, generally an ignored segment of the wastewater treatment problem. Concern with nutrient removal goes back within the last two decades to the realization by many, that fertilization of receiving waters by effluents, whether point source or nonpoint source, creates severe down stream problems, upsets the ecological balance, and can lead to detrimental impact on man's activities, but more importantly, his water resources. Thus the rising tide of action on effluent quality control, extending beyond removal of oxygen demanding organics. What this effort has represented is a recognition, however belatedly, that a cyclic system is involved, and that all segments of the system are of importance. This facet is critical to solution of water quality problems inherent in man's interaction with the environment.

It is noted that legal recognition of the importance of wastewater in the systems concept certainly occurred in the development of PL92-500[1], Federal Water Pollution Control Act Amendments, passed in 1972. The law's provisions set out not only the reason why topics such as nutrient removal must be considered, but also the required goals in attainment. For example, it is stated that there shall be achieved,

1. "....not later than July 1, 1977, effluent limitations for point sources....which shall require application of the best practicable control technology currently available...."
2. and, "....not later than July 1, 1983, effluent limitations for point sources,....which shall require application of the best available technology economically achievable...."
3. and, also, "....the national goal that discharge of pollutants into the navigable waters be eliminated by 1985."
4. continuing, "....the objective of the Act is to restore and maintain the chemical, physical and biological integrity of the Nation's waters,...."
5. further, "....water management methods applicable to point and nonpoint sources of pollutants to eliminate discharge...."
6. and, "....develop, refine, and achieve practical application of....advanced water treatment methods applicable to point and nonpoint sources...."
7. lastly, "To the extent practicable, waste treatment management shall be on an area wide basis and provide control or treatment of all point and nonpoint sources of pollution...."

As is evidenced by the foregoing, the system requirements are extensive, including both point and nonpoint sources, escalating process performance with time, culminating with elimination of pollutant discharges by 1985, consideration of chemical, physical, and biological parameters, advanced waste treatment, and lastly, consideration of best available technology economically achievable. All of these things compound into a stiff task for the design engineer. And this task is constrained by and dependent on the effluent limitations set by regulatory authority under the law.

Note that under the Clean Water Act of 1977 the actions of PL92-500 have received some fine tuning corrections, for example, redefinition of time schedules, but the basic tasks remain.

It is instructive now to examine the system fraction associated with point sources.

WATER RESOURCE SYSTEM

For many years, because of plentiful water, smaller population, and less technological impact, water supplies and effluent disposal problems were separately evaluated and were not really considered connected. Information now extant on trace chemicals in water, production of chloroform, and similar, finally have demonstrated that the water resource system must be examined, rather than its isolated parts. This system can be described as shown in Figure 1. Water taken from a source

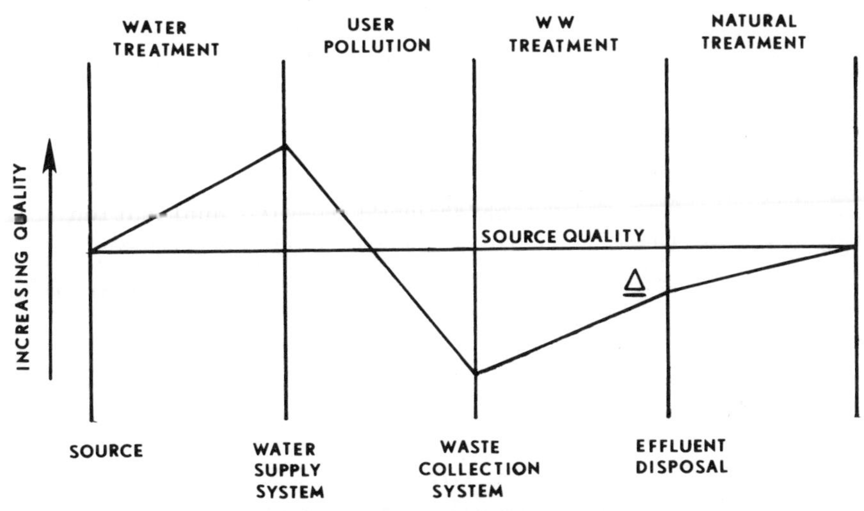

FIGURE 1. WATER RESOURCE SYSTEM
Note: Adapted from McLellon, et al. (7)

is improved in quality, delivered, degraded, collected, improved through wastewater treatment, then polished by the natural environment. A delta function exists, which in the past has been extensive. This delta function gradually is being reduced with the improvement of old wastewater treatment plants and construction of new ones. Basically, within the 1977 goal of PL92-500, the delta function amounted to the residuals from secondary treatment, with 90% BOD reduction, leaving residual nutrients of nitrogen and phosphorus, along with the enrichment resulting from the processing of remaining BOD by the natural environment.

The design engineer in this less complex case had available to him several process selections, such as the activated sludge variants or trickling filters, along with solids separation techniques, to adequately handle most problems. Process selection in many cases was routine. Later developments have included the biodisk surface contactor technique, aerated lagoon variants, and similar applications of biological and physical treatments.

This system interrelation has become more stringent as the complexities of modern chemical technology have impacted. Thus the objective of reducing the undesirable parts of the delta function, i.e., pollutants, to zero discharge by 1985. Realistically this cannot include similar reduction of the mineralization by soluble items such as sodium or chlorine but must be restricted to such things as the nitrogen, phosphorus, and BOD, which have immediate serious impact. It could be extended to treatment produced pollutants, such as the chlorinated compounds, which again impact on the process design. The task of the design engineer simply becomes harder and he must have greater knowledge and understanding of the biology and chemistry of design, and their process interactions. Increasingly also, more careful attention will need to be paid to the detailed measurable quality parameters both before and after treatment. It is useful here to consider the general design problem.

GENERAL DESIGN PROBLEM

The design engineer is faced with a quality conversion necessity in treatment, usually operating at a high hydraulic rate. The problem might be reduced to the simple diagram of Figure 2. In the past much standardized design for municipal

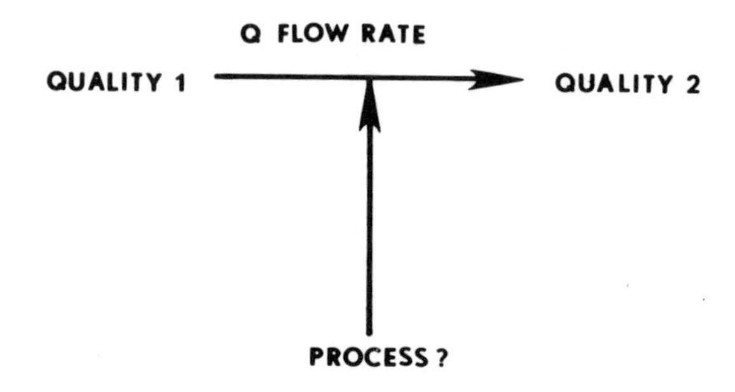

FIGURE 2. GENERAL DESIGN PROBLEM

waste treatment was done to a large extent on the basis of hydraulics, witness the design either of trickling filters or activated sludge processes for general reduction of BOD. BOD was recognized as a gross biological loading in design calculations. Over the past decade or so however, considerably more attention has been focused on the process details and the reasons for, or methods of, making quality changes. This has been spurred by recognition of downstream effects, hence studies of fate of pollutants have been emphasized, along with increasing application of improved instrumentation to determine parameter levels at critical points in the treatment and disposal process.

Note that in the general design problem, the engineer must consider the following:

Quality 1 - Unless some type of regulatory control exists upstream, the treatment plant to be designed must accept all quality variants recognized in an annual flow. But beyond this the plant must be prepared to treat, with some possible process improvements, of course, over its life span of possibly 20 to 30 years. Thus the design engineer doing the process design should, of necessity, have detailed quality information of high reliability, with projections if practicable.

Quality 2 - Acceptable levels of quality for critical parameters are those set by regulatory authority based on the best available scientific and engineering advice. Note that the advanced waste treatment standards published in Florida are 5-mg/l BOD, 5-mg/l SS, 3-mg/l N, and 1-mg/l P at present. These published levels provide guidance, but one should recall the national goal in PL92-500. It gives as a goal zero discharge of pollutants by 1985. For the design engineer there is a world of difference in the two, i.e., the difference between zero and one, for example. Hence, the Quality 2 parameters for design can be only those that are stipulated by some regulatory function, based on future expectations, where critical items such as nitrogen are involved. It is not possible technically or economically to say zero; instead, some value different from zero must be used that is reasonably attainable. As indicated by Barth,[2]

> "....no matter what process or sequence of treatment processes is applied to municipal wastewater there remains a residual of 1- to 2-mg/l of soluble organic nitrogen in these effluents."

Barth also notes this residual is not removed well by activated carbon and it is resistant both to chemical and biological oxidation. This particular nutrient also is received naturally in rainwater; thus, a low, realistic level is indicated, not zero.

Process - What is wanted is a process train that will solve the known problem, i.e., conversion from quality one to quality two for each parameter at risk. But beyond this the process should be adaptable and flexible, in view of the fact that it will receive currently unknown but estimated insults and, additionally, must handle the quality challenges of many years hence. This process train will be a combination of physical, chemical, and biological operations and processes in most cases. Note that it can be extremely varied, extending as it does from closely controlled processes such as those of a restricted treatment plant site to natural environmental segments included in a process train. An example of the latter could be a series of land disposal steps.

Now this general design process can be applied to the case at hand, i.e., the Biological Nutrient Removal Alternatives. It is of interest to consider first the types of cases.

NUTRIENT REMOVAL CASES

Since large investments currently exist in operating wastewater treatment plants, it is apparent that several different conditions occur in the nutrient removal field. The first is an existing, modern secondary plant requiring upgrading to tertiary treatment standards. A second is the new plant design replacing several currently inadequate plants, perhaps on a regional basis. Cases also exist of incorporation of these plants into an upgraded plant scheme, as differentiated from simply tacking on an application of tertiary treatment to accomplish nutrient removal. These cases apply primarily to point sources of pollutants. Lastly is the application of treatment techniques to nonpoint sources. The latter sources are identified in discharge location in many cases but are disperse and of large number, hence their collection into large process streams is difficult and expensive.

Now in the new plant, point source case, the process design naturally can be wide ranging, commencing with the Quality One input from the delivery sewer. Note at this point, some substantial fraction of the nitrogen and phosphorus would be present in the suspended solids and the remainder generally as ammonia in solution. As an example, for medium strength domestic sewage, Metcalf and Eddy[3] list 15-mg/l organic nitrogen and 25-mg/l ammonia, with nitrite and nitrate, the oxidized forms, absent. Organic phosphorus is listed as 3-mg/l and inorganic as 7-mg/l. Nutrient removal commencing at this point extends to the designer's options in suspended solids removal as well as conversion in chemical form as an intermediate step. Thus a complete chemical-physical system could be devised, considering the parameters existing. The designer would have the widest latitude of choice with the constraint of reduction of BOD/COD and similar. With a selection of initial biological conversion instead, unless complete nitrification occurs, the tertiary treatment problem shifts somewhat because of the reduced carbon source present after the biological treatment step. The BOD has been removed. In either case, the process selection must proceed in the same fashion as the general design case, i.e., consideration of

- Quality 1 - In detail, with the forms and levels of pollutants present, particularly BOD, N, P, SS, etc.
- Quality 2 - The objective, in detail, which depends on the use of the effluent and the regulatory climate, currently tertiary treatment standards for BOD, SS, N, P.
- Process - Alternative process trains for accomplishing the result are developed.
- Economics - The feasibility of accomplishment of each alternative is investigated.

Beyond these I believe there are two points affecting the design which need further consideration and often are not mentioned. They are

Instrumentation and Control - Whether sophisticated, automated, etc.

Operating Capability of Client - What reasonably can be accomplished in operation by the client agency and how does this impact the design?

The latter two items are important, in my opinion, from observation of plants and trends. Our technology is advancing rapidly. It is not apparent to me that our operating capability is advancing in like fashion, so this certainly affects the process design.

The other alternative cases would be approached in the same way as the preceding, commencing with detailed analysis of the end points and identification of process selection alternatives. It is noted though, that some additional complications have developed over recent years. That is, in addition to protecting the environment through effluent quality control, increasing emphasis is being placed on use of the environment as the whole or as a part of a treatment plant. Mr. Costle, EPA Administrator, recently reemphasized the EPA desire for land disposal to be a major alternative considered. What this amounts to is the designer applying agricultural (farm) and chemical process techniques to the treatment problem. It also adds to the need for knowledge, aimed at providing design guidelines, criteria, and standards, of the pathways and fate of pollutants in natural systems of various types. In Florida this could include intense agriculture, citrus, pasture, swamps, and other land variants. Data on pathways, fate of pollutants, and derived design criteria currently exist only in part for the U. S. Much work remains to be done because of large variation in physical, chemical, and biological features of topography, and the fact that releases to environments become uncontrolled in such cases, compared to a treatment plant flow. Thus, another requirement is placed on the designer because of this. In as complete a set as possible, data are needed on the site conditions - geology, hydrography, soils, etc., projected cover crop - trees, grains, grasses, etc., and lastly the operating capability. In this case an organized farm may exist where the operation cannot be neglected, just as with a treatment plant. As one other point, surveillance techniques, part of the design, must be developed. Reiterating, also, for any release, such as to a cypress bayhead, where final disposal is uncontrolled, good information is needed on the pathways and fate of pollutants, otherwise it is no different than a lake or river discharge with unknown results. Design implies producing a known end instead, with consideration of the over-all system.

In all of these cases, it is obvious that the quality change by a selected process train is the important facet. The quantity of flow is secondary, though the quality conversion process must function properly at all ranges of flow. The process must be reliable and flexible throughout its life, which places stringent demands on all segments handling the flow and producing the quality changes.

GUIDANCE

It is refreshing to note the large volume of information which is developing on treatment processes, particularly those for nutrient removal. The literature is replete with examples such as the "Process Design Manual for Phosphorus Removal", EPA 625/1-76-001a, April, 1976 and similar, under the Technology Transfer Program. The literature is being extended into areas of uncertainty. A critical example of the latter is the use of natural topography such as a cypress swamp or other natural environment for physical, chemical, and biological treatment, with uncontrolled end point. Recent proposals in the Kissimmee River basin, leading to restoration, and prevention of nutrient loading of Lake Okeechobee, provide other examples. Investigative efforts are planned in that basin to examine effects of such things as shallow ponding of nonpoint source runoff, plant growths, and delay structures.

NONPOINT SOURCE RUNOFF

Since it is unlikely that most wastewaters of this type will be treated by conventional treatment plants in any reasonable future time, treatment to reduce nutrient impact must rely on minimal structural corrective measures such as percolation or detention basins, then uncontrolled natural treatment and disposal. This amounts to land disposal of effluents from diverse locations using delay, plant growth, soil filtration, adsorption, and similar, to reduce nutrient impact. This work is in its infancy but is related to the land disposal of treatment plant effluents. Again, the unknown for the designer is the fate of pollutants in such things as swales and other commonly used structures, along with end points. The design problem is more uncertain than that of a point source, pending accumulation of additional information on the performance of corrective measures. Thus it is not possible to design to a Quality Two end result with certainty, but only at this time to employ land disposal measures which will improve effluents. As time passes experience should add to specific knowledge and allow a better predictive design. The requirements of PL92-500 for application of management techniques to nonpoint sources can be effected without stipulation of a fixed end result.

OPERATING CAPABILITY

Advanced wastewater treatment and nutrient removal in a point source plant add to the complexity of the treatment process. Since reliability and flexibility are needed to insure production of a fixed minimum quality result in the effluent, in the face of varying input, a requirement for stringent process control is placed on the plant. This has a concomitant requirement of adequate instrumentation, with timely, accurate, intelligible, and interpretable results, packaged into an analyzable form. Thus the instrumentation requirements may be formidable. But these requirements cannot be considered independently of the operating capability, i.e., of the client management and personnel of the plant, however much the design engineer would like to do so. The process design, physical facilities, instrumentation, operating personnel, and operating manual are intimately connected and this fact cannot be ignored. As just one example, consider the startup of a new plant, which is a complex chemical process industry equivalent. It has been estimated[4] that a sum equal to 5 to 15% of capital investment may be needed for startup. What is involved? A host of things are needed such as instrumentation checks and calibrations, equipment operating tests, pressure tests, and unit checks, extending into unit operation and process startup, then extending segment operation into plant operation and shakedown as a whole. Presumably the client agency provides the operating crew for startup and continuing operation. But note that while startup involves a mix of design personnel, manufacturer/supplier/contractor personnel, and client operations staff, the final continuing result depends on the client staff. In addition, an operating manual and conversion training are necessary to insure competent operation of the plant. The latter operating and training need must be extended into consideration of the number, types, and quality of operating staff and their continuing training program. Lastly is a check by design personnel at periodic intervals after startup to insure proper process operation.

It is unreasonable, in my opinion, to establish a process design, complex and sophisticated instrumentation level, and severe operational need, without very close cooperation with the client operating staff. A wastewater treatment facility which is publicly owned has a different operating capability from a plant of the chemical process industries. It is not intended to imply here that competent staffs cannot be attained. However, it is noted that the problem is more difficult and if it can be alleviated by alternatives in the design process well and good. Close contact is indicated between designer and operating personnel through the design process to insure a design that can be started up with minimum trauma, then can be operated reliably by the client operating staff to a successful end result. The designer has a stake in the latter to a greater extent than in the process design itself, which is worthless unless operated properly. Thus the designer, in establishing the design alternative to be used, must keep this operating need paramount.

COSTS

Nutrient removal is expensive, which means that the process alternatives must be thoroughly investigated. In particular, because of EPA guidance, wastewater treatment by land application will be an alternative which must be considered. It is noted that the EPA has issued guidance on costs in the 1976 report entitled "Costs of Wastewater Treatment by Land Application".[5] In that document, performance of land application is listed as 85 to 99% BOD and SS removal, to 90% nitrogen removal, and to 99% phosphorus removal. Bouwer, et al.,[6] in a recent 1978 paper, cite removal by land treatment as a viable alternative to secondary or tertiary plant treatment. Thus, since process performance of alternatives may be equal, cost will be one of the major deciding factors. The trickling filter and activated sludge processes have been main process techniques in earlier wastewater treatment. Now, with the extension into nutrient removal, with many more process variants, more stringent effluent limits, and complex technology, cost comparison of alternatives should become even more important to the designer. In this connection, it is believed that increasing need will exist for applied research and analysis on many problems, thus adding to the cost before process alternatives adequately can be assessed. This is so because of current uncertainties in technical solutions.

CLOSURE

The subject of process selection involves consideration of the influent and effluent quality parameters in detail. The rate of flow is a secondary factor which must be satisfied hydraulically, and where interrelated with the chemical or biological process, such as in recycling streams in a plant to maintain a culture concentration. The objective is to convert the quality at minimal cost, using a process train which is reliable, flexible, and capable of operation by client staff. Inexorably, process trains, instrumentation, and controls are growing more complex. This means that client operating staff must be of sufficient caliber and trained to satisfy these needs. The designer, in approaching the process design, related instrumentation, startup, and operation, must consider the client operation capability as a factor in process design decision. Lastly, with the growth in number of process alternatives, including the land application alternative of EPA interest, and the rising costs of technology, a most critical part of process selection will be preparation and use of valid, detailed cost estimates.

REFERENCES

1. "PL92-500, Federal Water Pollution Control Act Amendments of 1972," U.S. Federal Water Pollution Control Administration, U.S. Government Printing Office (1972).

2. Barth, E.F. Implementation of Nitrogen Control Processes, Workshop on Biological Nutrient Removal Alternatives, Florida Technological University, Lake Buena Vista, FL (1978).

3. Metcalf and Eddy. Wastewater Engineering (McGraw-Hill Book Company, New York, 1972).

4. Gans, M. "The A to Z of Plant Startup," Chemical Engineering 83(6):72-82 (1976).

5. Pound, C.E., R.W. Crites, and D.A. Griffes. Costs of Wastewater Treatment by Land Application, Technical Report, EPA-430/9-75-003 (1976).

6. Bouwer, H., W.J. Bauer, and F.D. Dryden. "Land Treatment of Wastewater in Today's Society," Civil Engineering 48(1): 78-81 (1978).

7. McLellon, W.M., W.M. Drewry, and J.M. Glennon. Correspondence Course Manual for Water Plant Operators, Class C, 304 p, The South Carolina Water and Pollution Control Association and Clemson University, Clemson, SC (1971).

2. IMPLEMENTATION OF NITROGEN CONTROL PROCESSES

E. F. Barth, U.S. EPA, Office of Research and Development, Municipal Environmental Research Laboratory Cincinnati, Ohio

ABSTRACT

Control of the nitrogenous residual content of municipal wastewater effluents is a complex problem. Applied technique from the sciences of microbiology and chemistry, as well as detailed engineering considerations must be blended for process stability. Highly efficient operation of nutrient control processes has been demonstrated. Contemporary experience can be utilized to guide future designs and regulatory decisions.

INTRODUCTION

This work on biological nutrient removal alternatives takes place at a pivotal time in the national goal of elimination of pollutants from wastewater discharges. About 10 years ago the first full scale nutrient control studies were completed and rapid implementation of this technology took place. It can be foreseen that this effort will continue into the next 15 years as new construction occurs. These workshop discussions should therefore prove useful to current planners of new facilities. Figure 1 shows the progress of nutrient control technology.

The program agenda shows very pragmatic workshop type presentations, however since the major sponsor of the program is a university it would not be out of line to deviate briefly into slightly academic topics.

From the viewpoint of environmental control it should be asked: What is a nutrient, and what are the nutrients? The first definition must be addressed from a basic biochemical view. A nutrient is a substance (in association

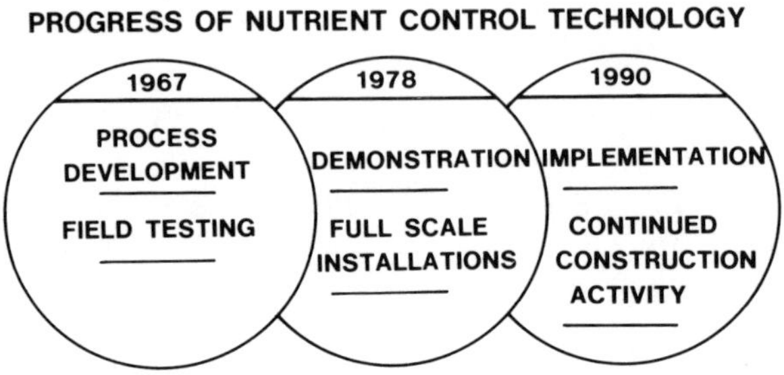

PROGRESS OF NUTRIENT CONTROL TECHNOLOGY

1967
PROCESS DEVELOPMENT
FIELD TESTING

1978
DEMONSTRATION
FULL SCALE INSTALLATIONS

1990
IMPLEMENTATION
CONTINUED CONSTRUCTION ACTIVITY

FIGURE 1. PROGRESS OF NUTRIENT CONTROL

with other substances) that provides for growth and reproduction of an organism. These substances have been variously identified as complex materials like carbohydrates, amino acids and specific chemical elements. In the environmental control field only the chemical elements are addressed as nutrients.

From an elementary standpoint the nutrients that are generally regarded as essential to organisms can be listed by a professorial mnemonic:

C. HOPKINS CaFe

The letters represent the symbols of various chemical elements. Of these 10 elements, at the present time, there exists engineered systems for control of three - carbon, phosphorus and nitrogen - which will be the subjects of this two day workshop. If required, control of sulfur and iron could be accomplished. There is no current feasible process for the direct control of calcium, hydrogen, oxygen, potassium and iodine. Table I illustrates these points.

Autotrophic organisms such as algae require all these elements for growth during the cultural condition called eutrophication. Therefore from a biochemical viewpoint this workshop will consider only 30 percent of the nutrient problem. However very thorough investigations of eutrophic responses indicate that control of nitrogen and phosphorus (singly or in combination) can control eutrophication; therefore from a practical standpoint the workshop will address 100 percent of the solution to eutrophication.

TABLE I. Control Technology for Elements.

Control Technology	C.	H	O	P	K	I	N	S	CaFe
Available	X			X			X		
Possible								X	X
Not Feasible		X	X		X	X			X

The rest of this discussion will center on the basic chemistry and biology of nitrogen control, efficiency and reliability of installed processes and the implications of these on regulatory decisions concerning nitrogen in effluents.

NUTRIENTS IN WASTEWATER

The cultural and societal structures of the United States are rather fluid and this can be reflected in the composition of municipal wastewater. The phosphorus content of most municipal wastewaters has declined from 12 to 10 mg/l to levels of 5 to 3 mg/l over the last five years. The lower values being in areas that have instituted bans on phosphate laundry detergents, and a general reduction in the phosphate content of most packaged detergents. This causes some minor design changes and certain savings in operational costs for phosphorus control processes. It does not eliminate the need for some form of chemical control. Even at these lower influent phosphorus concentrations municipal wastewater is still deficient in carbon with respect to the heterotrophic metabolism of the organisms utilized for biological treatment. The low effluent residual phosphorus permitted in eutrophic cases and the high reliability of target attainment still necessitates chemical control.

Nitrogenous components of wastewater are not so sensitive to cultural and societal change. The major nitrogenous fractions of wastewater are the result of human metabolism and only a drastic change in diet or physical modification of toilet fixtures would alter the concentration. Therefore basic designs for nitrogen control, necessary chemical additions or managed biological systems will change very little in the foreseeable future.

One other major difference between design consideration for nitrogen and phosphorus needs to be mentioned. In most biological environments, and in all wastewater treatment processes, phosphorus exists as phosphate and does not change chemical valence. Nitrogen on the other hand cycles between several chemical valence changes. Treatment plant design must account for these various specific forms of nitrogen. Control of nitrogen in effluents is much more complex and difficult than phosphorus control. Table II shows examples of chemical changes during biological treatment.

TABLE II. Cycles of Nutrients.

Element	Biological Process	
	Oxidation	Reduction
Carbon	CO_2	CH_4
Oxygen	CO_2	H_2O
Hydrogen	H_2O	H_2
Sulfur	$SO_4^=$	H_2S
Nitrogen	NO_2^-	Org-N
	NO_3^-	N_2, NH_3
Phosphorus	Remains as $PO_4^{\equiv}$ in both systems	

The environmental concerns with wastewater nitrogen are multi-faceted. Three major concerns can be identified as health, oxygen demand and eutrophication. These can be ranked according to various social parameters as shown on Table III. Human health would have to be rated more important than other aspects. Oxygen demand has the greatest impact because more facilities and a greater volume of wastewater require this type control than the other two problems. Health could be protected with 100 percent success by furnishing oxidized-nitrogen-free water to infants. Eutrophication control is most costly because in this circumstance a nitrogen removal process must be implemented, rather than controlling the form of nitrogen discharged or limiting ingestion of suspect water. There are other concerns with the environmental fate of nitrogen but these serve to illustrate the complex nature of the subject.

INTERACTION OF ANALYTICAL IMPLICATIONS AND REGULATIONS

As previously mentioned the chemistry of nitrogen is complex. To design, operate and evaluate a nitrogen control process the various forms of nitrogen such as organic nitrogen, oxidized nitrogen and ammonium nitrogen must be accounted for. In the case of a facility that was required to achieve total nitrogen removal, to some stated residual by a regulatory decree, all these nitrogen forms would have to be summed.

TABLE III. Environmental Concerns of Nitrogen Manifestations.

Social Topic	Concern In Decreasing Order
Importance	Health Oxygen Demand Eutrophication
Impact	Oxygen Demand Eutrophication Health
Ability to Control	Health Oxygen Demand Eutrophication
Cost of Control	Eutrophication Oxygen Demand Health

Table IV gives the approximate precision of these analytical tests as reported in the 14th Edition of Standard Methods. It is evident that at the 1 mg/l level the test results can be highly variable; and summation of the various forms of nitrogen could produce an uncertain value.

Recent investigations of nitrified effluent have rediscovered an analytical problem known for a long time by chemists in the business of analyzing fertilizer. When the ratio of nitrate nitrogen to organic nitrogen in a sample exceeds 10:1 low recoveries of organic nitrogen are found by the Kjeldahl procedure. Under these conditions partial loss of nitrogen occurs as nitrous oxide gas during the digestion process.

Detailed studies of many secondary, advanced and tertiary effluents have shown that no matter what process or sequence of treatment processes are applied to municipal wastewater there remains a residual of 1 to 2 mg/l of soluble organic nitrogen in these effluents. The material is resistant to chemical oxidation by chlorine, ozone and permanganate; as well as biological oxidation. Activated carbon does not adsorb the material very efficiently. Table V provides data from several treatment plants.

It should be noted that the substrate concentrations of ammonium nitrogen and nitrate nitrogen that start to limit the rate of reaction in nitrification and denitrification are about 0.5 and 1 mg/l respectively. Regulations requiring levels lower than these values would require unrealistic reactor detention times, resulting in escaladed

TABLE IV. Approximate Precision of Various Nitrogen Analyses at ∿ 1 mg/l Level.

Analysis	Relative Standard Deviation, Percent
Ammonia Nitrogen (distillation + titration)	±21
Nitrate Nitrogen (cadmium reduction)	±25
Nitrite Nitrogen (diazotization)	(not given)
Organic Nitrogen (titration)	±45

TABLE V. Soluble Organic Nitrogen in Wastewater.

		Soluble Organic Nitrogen, mg/l		
Location	Sample	Inf.	Pri.	Sec.
Union City, CA	1	4.1	3.5	1.1
	2	6.3	4.9	1.9
Palo Alto, CA	1	5.5	4.4	2.1
	2	4.9	4.4	1.2
San Jose, CA	1	4.6	4.1	1.5
	2	6.3	5.5	1.8
Tahoe, CA	1	2.9	-	1.2

cost. On the other hand nitrogen control designs must be based on achieving essentially zero residual because the controlling factor of success in these processes is the net growth rate of microorganisms. A design based on some intermediate growth rate to achieve an intermediate nitrogen residual level would be unstable.

Due to rate limiting concentration, analytical variability, internal recycle streams, excursions in hydraulic and mass flow in real world operations; a realistic regulation for ammonium nitrogen concentration should not be lower than 1 mg/l. These same considerations, plus the inevitable soluble organic nitrogen, indicate that a total nitrogen regulation lower than 3 mg/l could not be consistently achieved by any cost effective technology known today.

These comments lead to the conclusion that regulations

should not be based on never-to-exceed values. A realistic regulation would be based on probability of occurrence of a stated value. Figure 2 illustrates a probability plot of effluent total nitrogen from a system utilizing two-stage trickling filters followed by denitrification on a mixed media filter using methanol addition. Fifty percent of the observations were 2.5 mg/l or less. However values as high as 5 mg/l and as low as 0.5 mg/l can be encountered as extremes.

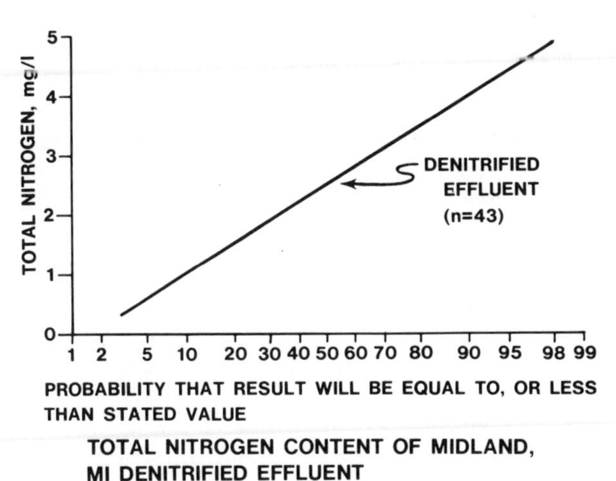

FIGURE 2. EFFLUENT TOTAL NITROGEN PLOT

CURRENT TECHNOLOGY FOR NITROGEN CONTROL

Experience has culminated in resolving the choice between physical-chemical and biological designs for nitrogen control. Biological systems have proven clear over all superiority.

Physical-chemical systems have clear limitations. Most of all these type processes must operate as nitrogen removal systems, whereas a biological system can be designed to control the form of nitrogen discharged, if that is a suitable solution. Therefore a physical-chemical selection would be more costly from a general viewpoint.

Since nitrogen exists in different chemical forms in wastewater it is difficult to design a highly efficient physical-chemical process for total nitrogen control. Biological processes can be managed to convert the nitrogen to one predominant chemical species.

Several physical-chemical processes utilize high concentrations of chlorine for removal of nitrogen. The present day awareness of the environmental effects of chlorine and chlorinated materials argues against application of this technique.

Operational characteristics for biological processes are simpler and less dynamic than operation of physical-chemical processes. This leads to less maintenance requirements and more stable results.

Currently there are about 200 nitrogen control facilities in operation, ranging in size from 1,000 m^3/d to 1,140,000 m^3/d. The vast majority are biological nitrification systems. Only a few total nitrogen removal facilities are presently in operation.

IMPLEMENTATION OF NITROGEN CONTROL PROCESSES

In April 1977 EPA's Office of Water Programs Operations reviewed the 1976 Needs Survey, interviewed all 10 EPA Regional Offices and reviewed pending construction grant projects in regard to use of new technology. Some of the conclusions are of direct interest to this workshop.

By the year 1990, 34 percent of the U.S. population will require secondary treatment, 55 percent will require greater than secondary quality effluent and 11 percent of the population will still be unsewered.

From the Needs Survey the projected number of plants that will eventually have to achieve various nitrogen residuals in their effluent can be abstracted. These are tabulated in Table VI. Oxygen demand is the dominant factor in this projection as evidenced by the number of plants that must meet an ammonium nitrogen limitation. The number of anticipated nitrogen removal facilities is only 15 percent of the number of nitrification facilities.

The survey of the various EPA Regions revealed that several design options were being utilized for nitrification. Table VII lists these options and the respective ranking of the options for construction grant applications present in the Regional Offices on the date of the survey (November 1976). Slightly more than half the options employ a two-stage process. Thirty percent of the options employ attached growth processes such as biological discs or biological filters, the rest are suspended growth systems. Systems employing oxygen, instead of atmospheric air, were 13 percent of the total.

CONCLUSIONS

Implementation of nitrogen control at municipal facilities is a complex design situation due to the interaction of biology, chemistry and engineering. A clear cut need for oxygen demand control is evident. Facilities already in operation have demonstrated that efficient nitrogen control can be obtained with proper design and operational control. This contemporary experience can be utilized to guide the projected 3,000 nitrogen control facilities anticipated by the year 1980.

TABLE VI. Effluent Nitrogen Residuals, mg/1 From 1976 Needs Survey.

TN In Effluent	# Of Plants	NH_4-N In Effluent	# Of Plants
<1	3	<1	89
1 to 4.9	198	1 to 1.9	1098
5 to 9.9	110	2 to 5	1479
			2666 Total
10 to 15	77		
	388 Total		

TABLE VII. Design Options for Nitrification Revealed By EPA Regional Survey.

Nitrification Option	Number Of Applications	Percent Of Total
Two-stage, air	60	42
Rotating biological disc	39	27
Two-stage, oxygen	16	12
Oxidation ditch	11	8
Single-stage, air	10	7
Activated biological filter	5	3
Single-stage, oxygen	1	1
	142 Total	100

3. KINETICS OF NITROGEN REMOVAL FOR MUNICIPAL AND INDUSTRIAL APPLICATIONS

W. Wesley Eckenfelder, Jr., Environmental and Water Resources Engineering, Vanderbilt University, Nashville, Tennessee

Yerachmiel Argaman, Technion-Israel Institute of Technology, Haifa, Israel

INTRODUCTION

Biological nitrification and denitrification is still one of the most economical processes for nitrogen removal from wastewaters, both municipal and industrial. The design and economy of such systems are closely related to process kinetics, i.e., to the rates at which nitrification and denitrification proceed. These rates, in turn, depend on the microbial population, the composition and concentration of the wastewater, and a wide variety of chemical and physical parameters. For municipal applications the design parameters are relatively well established since domestic wastewater have a more or less uniform quality. Industrial wastewater on the contrary, will vary in quality from one case to the other and will therefore require development of design parameters for each specific case. This difference in wastewater quality may also lead to different strategies of nitrogen removal in the two sectors.

Complete removal of nitrogen from wastewater can be accomplished in either a multistage system of reactors or in a recycled system. In the former configuration, commonly known as the "two sludge" or "three-sludge" system, BOD removal, nitrification and denitrification take place in separate activated sludge reactors. Recycled systems consist of a single reactor, divided into aerobic and anaerobic zones, with the same sludge being recycled through both. A basic feature of the recycled systems is the utilization of wastewater BOD as the hydrogen acceptor in the denitrification reaction. This may eliminate the need for an external carbon source and reduce the energy requirements for oxygen supply. Although for municipal applications this may be the system of choice, industrial wastewater may sometimes be more economically treated in the multistage system. This is particularly true when the wastewater can be segregated to several streams, each containing predominantly one of the pollutants, e.g. BOD, ammonia nitrogen, or nitrate nitrogen. Selection of the most economical system is facilitated by considering the kinetic principles as discussed herein.

NITRIFICATION KINETICS

Nitrification is the biological oxidation of ammonia to nitrate with nitrite formation as an intermediate. The microorganisms involved are the autotrophic species nitrosomonas and nitrobacter which carry out the reaction in two steps:

$$2NH_4^+ + 3\,O_2 \xrightarrow{\text{nitrosomonas}} 2NO_2^- + 2H_2O + 4H^+$$

$$2NO_2^- + O_2 \xrightarrow{\text{nitrobacter}} 2NO_3^-$$

The cell yield for nitrosomonas has been reported as 0.05 - 0.29 mg VSS/mg NH_3-N and for nitrobacter 0.02 - 0.08 mg VSS/mg NH_3-N. A value of 0.15 mg VSS/mg NH_3-N is usually used for design purposes.[1] The maximum specific growth rate of nitrosomonas is reported in the range of 0.01 - 0.05 mg VSS/mg VSS-hr, and that of nitrobacter averages 0.04 mg VSS/mg VSS-hr. It is generally accepted that the specific growth rate of nitrobacter is higher than the growth rate of nitrosomonas and hence there is no accumulation of nitrite in the process. Under these conditions the growth rate of nitrosomonas will control the overall reaction.

Since both ammonia nitrogen and molecular oxygen are required for the nitrification reaction, the rate of the process may be expressed by a Monod type equation:

$$R_N = R_{NM}\, X_{VN} \left(\frac{DO}{K_o + DO}\right)\left(\frac{N}{K_N + N}\right) \qquad (1)$$

where:
- R_N = rate of nitrification, mg NH_3-N/hr
- R_{NM} = maximum specific nitrification rate, mg NH_3-N/mg VSS-hr
- X_{VN} = concentration of nitrifying microorganisms, mg/l VSS
- DO = concentration of dissolved oxygen, mg/l
- N = concentration of NH_3-N, mg/l
- K_o = saturation constant for dissolved oxygen, mg/l
- K_N = saturation constant for ammonia nitrogen, mg/l

The reported value of K_o is approximately 1.0 mg/l and that of K_N is in the range of 0.2 - 1.0 mg/l. Hence, when both DO and ammonia concentrations are higher than 1.0 mg/l, nitrification approaches a zero order reaction. This was demonstrated by Huang and Hobson[2] who showed that the nitrification rate was essentially linear with biomass concentration even at very low nitrogen levels. Similar results were reported by Wong Chong and Loehr[3] for nitrogen concentrations ranging from 100 to 1200 mg/l. In general, the specific nitrification rate is obtained from the maximum growth rate and the yield coefficient:

$$R_{NM} = \frac{\mu_N}{a_N} \qquad (2)$$

where: μ_N = maximum growth rate of nitrifiers, mg VSS/mg VSS-hr

a_N = nitrifiers yield coefficient, mg VSS/mg NH_3^-N

Using the range of values for maximum growth rate and yield factors reported in the literature, the range of nitrification rate is estimated at 0.1 - 0.3 mg NH_3^-N/mg VSS-hr. Like all biochemical reactions the growth rate of nitrifiers is also temperature dependent.

This temperature dependence is expressed by:

$$\mu_{N(T)} = \mu_{N(10^oC)} \exp [0.11 \, (T-10)] \qquad (3)$$

where: $\mu_{N(T)}$ = specific nitrifiers growth rate at temperature T, mg VSS/mg VSS-hr

$\mu_{N(10^oC)}$ = specific nitrifiers growth rate at 10^oC, mg VSS/mg VSS-hr

T = temperature, ^oC

The reported value of $\mu_{N(20^o)}$ are in the range 0.02 - 0.09 mg VSS/mg VSS-day.

In most practical applications nitrification occurs in a mixed culture where both autotropic and heterotrophic organisms are present. The growth rate and yield coefficient of the heterotrophs are considerably higher than those of the autotrophs and in most cases the influent concentration of BOD is higher than that of ammonia. Hence, the rate at which the heterotrophs are produced greatly exceeds that of the autotrophs and the ratio of nitrifiers in the total biomass is given by:

$$f_N = \frac{X_{VN}}{X_V} = \frac{a_N N_{OX}}{aS_r + a_N N_{OX}} \qquad (4)$$

where: N_{OX} = ammonia nitrogen oxidized to nitrate, mg/l

a = heterotrophic microorganisms yield coefficient, mg VSS/mg BOD

In order to maintain an autotrophic population in a mixed culture of activated sludge, the systems sludge age must exceed the reciprocal of the autotrophs maximum specific growth rate. This was shown by Downing[4] and can be expressed mathematically as follows:[1]

$$\theta_c \geq 2.13 \exp[0.098\,(15-T)] \qquad (5)$$

where: θ_c = systems sludge age, days

It should be pointed out that equation 5 can be derived directly from equation 3 by letting $\theta_c \geq \frac{1}{\mu_N}$ and substituting $\mu_N(10^oC)$ = 0.012 mg VSS/mg VSS-hr.

The sludge age in the activated sludge process is given by:

$$\theta_c = \frac{X_v t}{a\,S_r - bxX_v t} \qquad (6)$$

where: t = hydraulic residence time, day
X_v = MLVSS, mg/l
a = yield coefficient for heterotrophs, mg VSS/mg BOD
S_r = BOD removal, mg/l
b = endogenous decay rate, mg VSS/mg VSS-day
x = biodegradable fraction of MLVSS

Equation 5 may be rearranged as follows:

$$t = \frac{A\,S_r\,\theta_c}{X_v(1+bx\theta_c)} \qquad (7)$$

Since X_v is limited in practice to approximately 3000 mg/l it is seen from equation 7 that the hydraulic retention time, t is proportional to the BOD removal and to the sludge age.

The BOD removal is expressed by:

$$\frac{S_r}{X_v t} = K_o\,\frac{S_e}{S_o} \qquad (8)$$

where: K_o = BOD removal rate coefficient, day^{-1}
S_e = effluent soluble BOD, mg/l
S_o = influent BOD, mg/l

Using equations 7 and 8 an optimized two stage system may be devised in which BOD removal is mainly limited to the first stage while nitrification is accomplished in the second. In many cases a single stage BOD removal and nitrification will be most economical.

Factors Affecting Nitrification Kinetics

Nitrifying organisms are subject to inhibition by various organic and inorganic compounds. Hockenbury and Grady[5] have summarized inhibition data for selected organic compounds. Anthonisen et al[6] have studied the effects of nitrogenous compounds on the rate of nitrification. They showed that inhibition results from free ammonia (FA) and free nitrous acid (FNA) the formation of both being dependent on pH and total concentrations of ammonia and nitrite nitrogen. FA inhibition to nitrosomonas begins at 10 to 150 mg/l, and to nitrobacter at 0.1 to 1.0 mg/l. FNA inhibition to nitrifying organisms initiated at concentrations between 0.22 and 2.8 mg/l.

The optimal pH for nitrification varies between 6 and 7.5, depending on the formation of FA and FNA. It should be noted from the nitrification reaction equations that 7.14 mg of alkalinity (as C_aCO_3) are destroyed for each mg/l of NH_3^-N oxidized. From equation 1 it is obvious that low DO levels will also affect the rate of nitrification. Hence, a DO level in excess of 2.0 mg/l is recommended in order to insure maximum nitrification rates. Heavy metals and many other organic and inorganic compounds are known to inhibit nitrification.

DENITRIFICATION KINETICS

Denitrification is the biological conversion of nitrate nitrogen to the more reduced forms such as N_2, N_2O and NO. The process is brought about by a variety of facultative heterotrophs which can utilize nitrate instead of oxygen as the final electron acceptor. It was shown that the breakdown of carbonaceous organics in the denitrification process is similar to that in the aerobic process, the only difference being in the final stages of electron transfer. Thus, the term anoxic denitrification would seem more appropriate than anaerobic denitrification. Thermodynamic data show a higher energy yield from the aerobic metabolism of organic carbon compared to anoxic denitrification. This would indicate the need for strict anoxic conditions in a denitrifying system. However, it was shown that under acidic pH conditions denitrification can take place in the presence of oxygen. Moreover, filmed reactors, as well as suspended growth systems, may consist of aerobic biomass layers and anaerobic sublayers so that aerobic processes and denitrification may occur simultaneously.

The stoichiometric reaction describing denitrification depends on the carbonaceous matter involved. Thus, for the most extensively used, and studied, external carbon source is methanol for which the reaction is:

$$6NO_3^- + 5\,CH_3OH \rightarrow 3N_2 + 5CO_2 + 7H_2O + 6OH^-$$

The theoretical methanol requirement is 2.47 mg MeOH/mg NO_3^-N. The equation predicts formation of 3.57 mg alkalinity for each mg of NO_3^-H denitrified. Field studies have shown 2.9 mg Alk/mgNO_3^-N.

Nitrate will also replace oxygen in the endogenous respiration reaction. The proposed equation is:[7]

$$C_5H_7NO_2 + 4NO_3^- \rightarrow 5CO_2 + NH_3 + 2N_2 + 6OH^-$$

The reported rates of endogenous denitrification around 20^oC vary between 0.9 - 2.0 mg NO_3^-N/g VSS-hr.[8 9 10 11 12]

The rate of denitrification can be derived from the growth rate of the denitrifying microorganisms. Since both carbonaceous matter and nitrate are involved the rate is expressed by a Monod type equation analogous to equation 1:

$$R_{DN} = R_{DNM} X_v \left(\frac{S}{K_s+S} \right) \left(\frac{NO}{K_{DN}+NO} \right) \qquad (9)$$

where:
R_{DN} = rate of denitrification, mg NO_3^-N/hr
R_{DNM} = maximum specific denitrification rate, mg NO_3^-N/mg VSS-hr
S = concentration of carbonaceous organic matter, mg/l
NO = concentration of nitrate nitrogen, mg/l
K_s = saturation constant for carbonaceous matter, mg/l
K_{DN} = saturation constant for nitrate, mg/l

The reported values of K_{DN} are below 0.15 mg/l, which indicate a zero order reaction with respect to nitrate down to very low nitrate concentrations. Values of K_s depend on the specific compound used as a carbon source. Reported values of the specific denitrification rates for various carbon sources at 20°C are summarized in Table I.

TABLE I. DENITRIFICATION RATES WITH VARIOUS CARBON SOURCES

Carbon Source	Denitrification Rate ($mgNO_3^-N$/mg VSS-day)	Ref.
Brewery wastes	0.22 - 0.25	16
Methanol	0.36 - 0.60	17
Methanol	0.15 - 0.40	1
Volatile acids	0.36	18
Glucose	0.06 - 0.07	19
Sewage	0.034-0.06	19
Sewage	0.05 - 0.07	7
Sewage	0.07	1

The BOD removal in the denitrification process follows the same relationship as in the aerobic process as expressed in equation 8. Experimental data derived from petrochemical wastewater treatability studies validate this, as seen in Figure 1. Wuhrmann[8] indicated that both the nitrate and oxygen respiration rates under endogenous conditions depend on the type and quantity of substrate which can be mobilized by the cells and therefore the two rates should be correlated. This is clearly seen from experimental data presented in Figure 2. The data also indicates that 1 mg of NO_3^-N is equivalent to approximately 3 mg of O_2. This is in good agreement with the theoretical value of 2.86 mg NO_3^-N/mg O_2.

Since the nitrate serves as a substitute for oxygen its rate of consumption may be expressed in a similar way. Thus, nitrate utilization in the biological breakdown of organic matter can be written as follows:

$$N_r = A'S_r + B'X_vt \qquad (10)$$

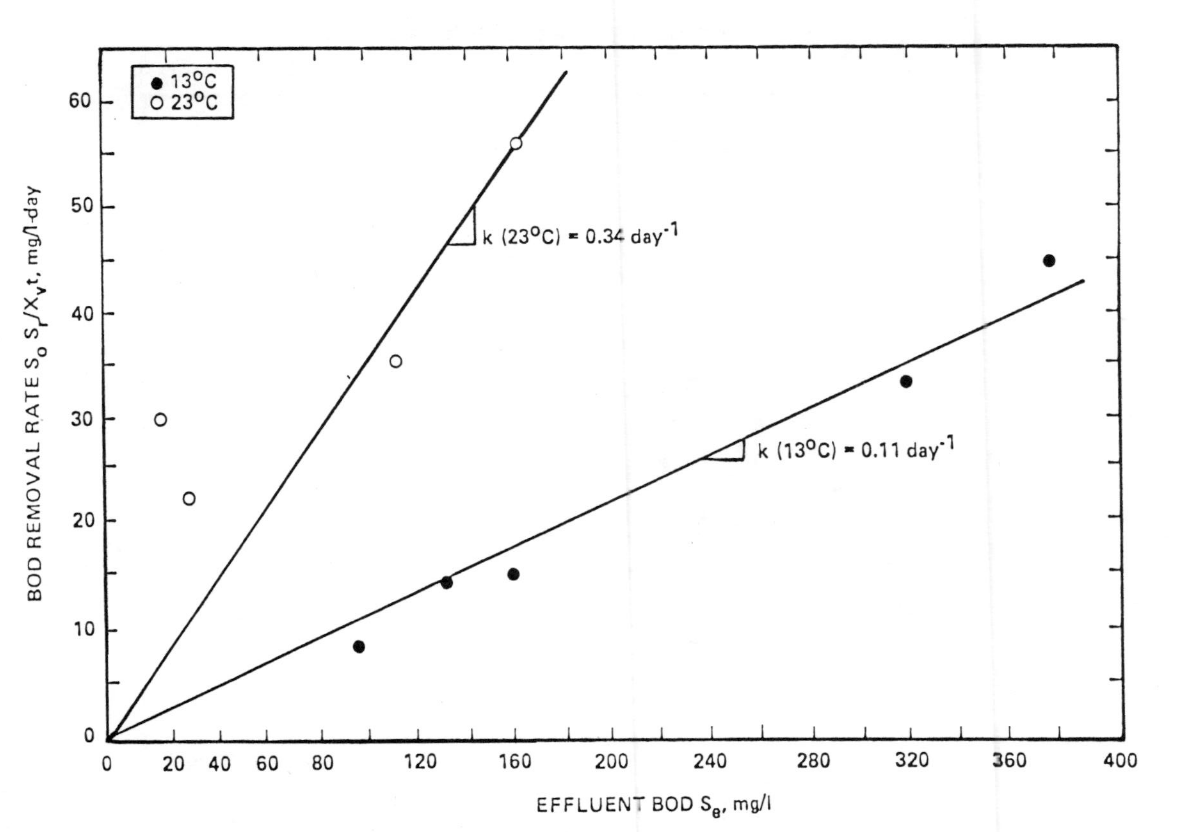

FIGURE 1. BOD REMOVAL RATE IN DENITRIFICATION

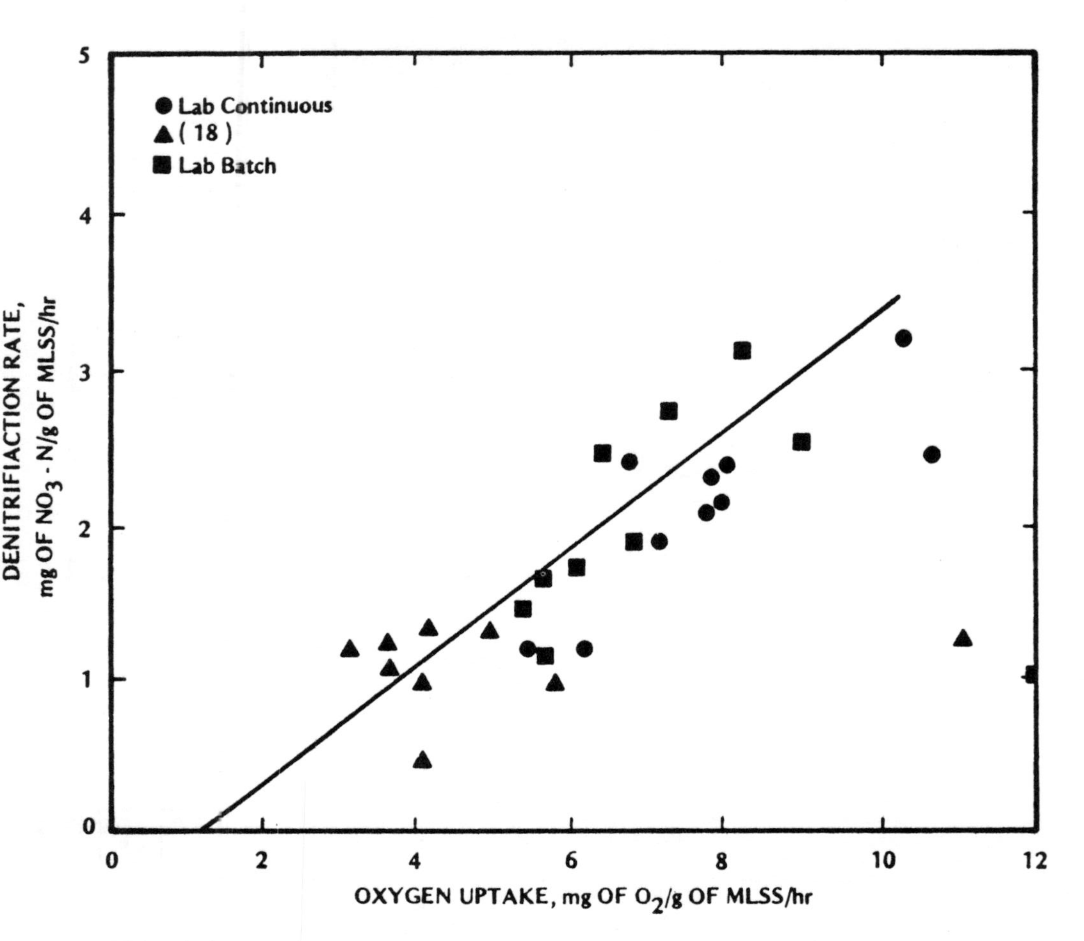

FIGURE 2. CORRELATION BETWEEN OXYGEN UPTAKE AND DENITRIFICATION

where: N_r = denitrified NO_3^-N, mg/l
A' = nitrate utilization for BOD removal, mg NO_3^-N/mg BOD
B' = nitrate utilization for endogenous respiration, mg NO_3^-N/mg VSS-day

Figure 3 illustrates experimental data from petrochemical wastewater studies, based on a rearranged form of equation 10. The coefficient A' and B' derived from this figure are 0.2 mg NO_3^-N/mg BOD, and 0.003 mg NO_3^-N/mg VSS-day, respectively. Recalling the theoretical relation between oxygen and nitrate these values are in good agreement with oxygen utilization coefficients.

The rate of denitrification varies with temperature as defined by the equation:

$$R_{DN(T)} = R_{DN(20^oC)}\theta^{T-20} \tag{11}$$

Values of θ for suspended growth systems vary from 1.07 to 1.20. Data for denitrification of an industrial wastewater is shown in Figure 4 from which the value of θ is calculated as 1.09. From the data of Figure 1 the calculated value of θ is 1.12.

RECYCLED SYSTEMS

Present economics mitigate against the use of methanol as a carbon source for denitrification. Rather, when treating domestic wastewater a recycled system of aerobic and anoxic zones is the most cost effective approach in most cases. Various forms of the recycled systems were described in the literature recently.[13 14] A mathematical model describing the process was presented by Argaman and Miller. The model can be used for process design under various conditions and treatment objectives. In contrast to the multistage system, it was shown that the influent composition affects the required sludge age. It is also closely related to the aerobic volume fraction as shown in Figure 5.

Apart from the elimination of methanol consumption the system offers savings in other operational cost items. Aeration power requirements may be reduced as part of the BOD is removed in the anoxic zone. Each mg of nitrate nitrogen converted to nitrogen gas is equivalent to 2.86 mg of oxygen. Chemical requirements for pH control are also reduced since part of the alkalinity destroyed in the nitrification stage is recovered in the denitrification reaction. These savings are partially offset by the increased energy used for mixed liquor recycling.

The design of a recycled system is somewhat more complicated compared to the multistage system. A trial and error technique proposed by Barnard[13] or a computerized solution as described by Argaman and Miller[15] should be used.

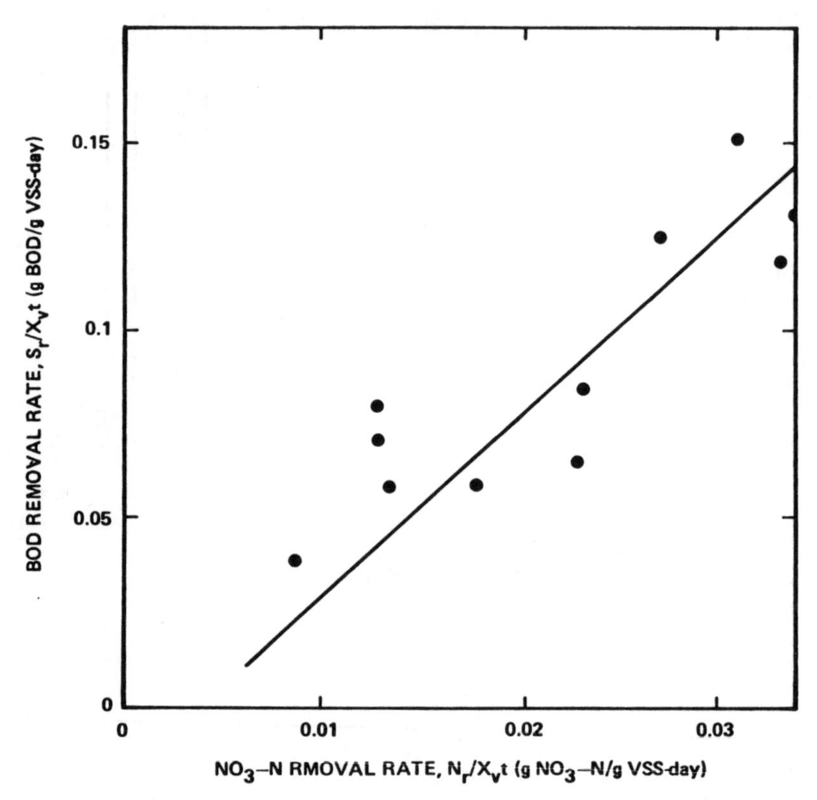

FIGURE 3. NO_3-N/BOD REMOVAL RATIO IN DENITRIFICATION

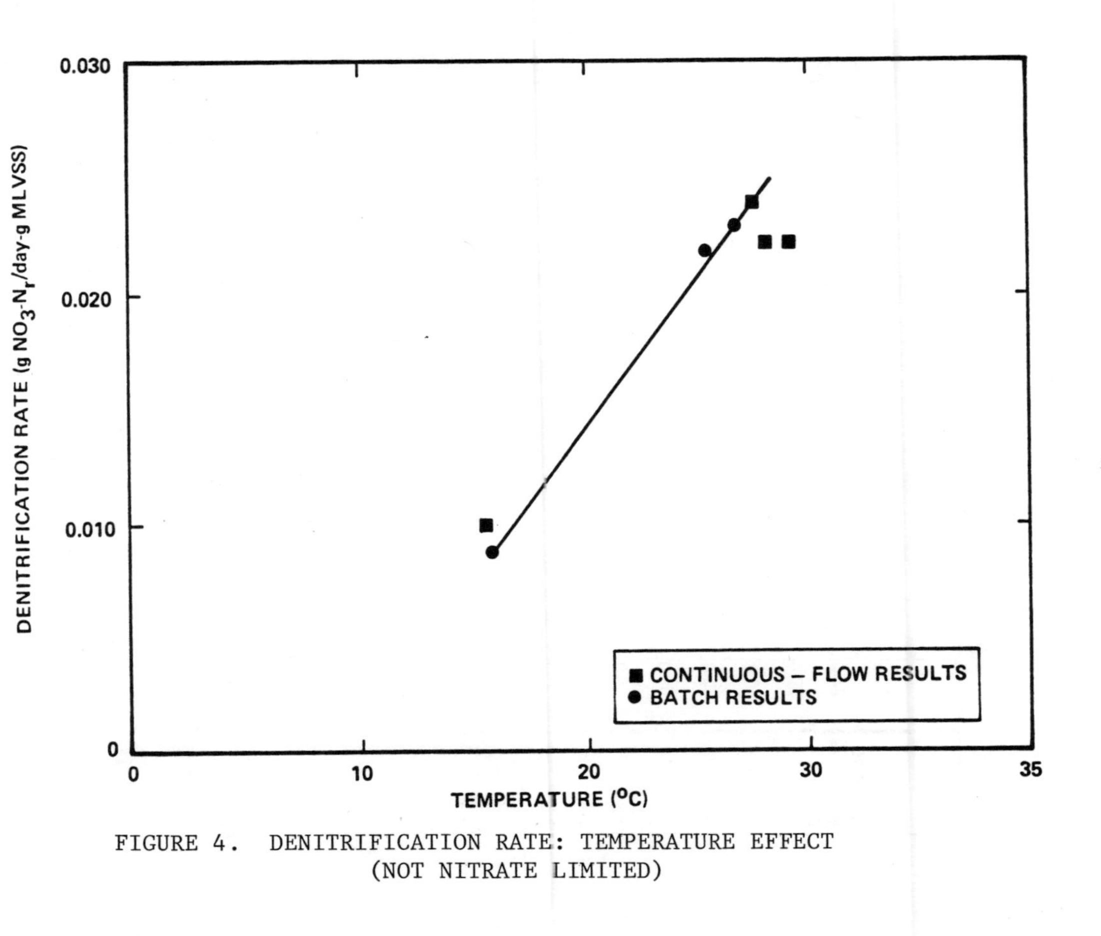

FIGURE 4. DENITRIFICATION RATE: TEMPERATURE EFFECT (NOT NITRATE LIMITED)

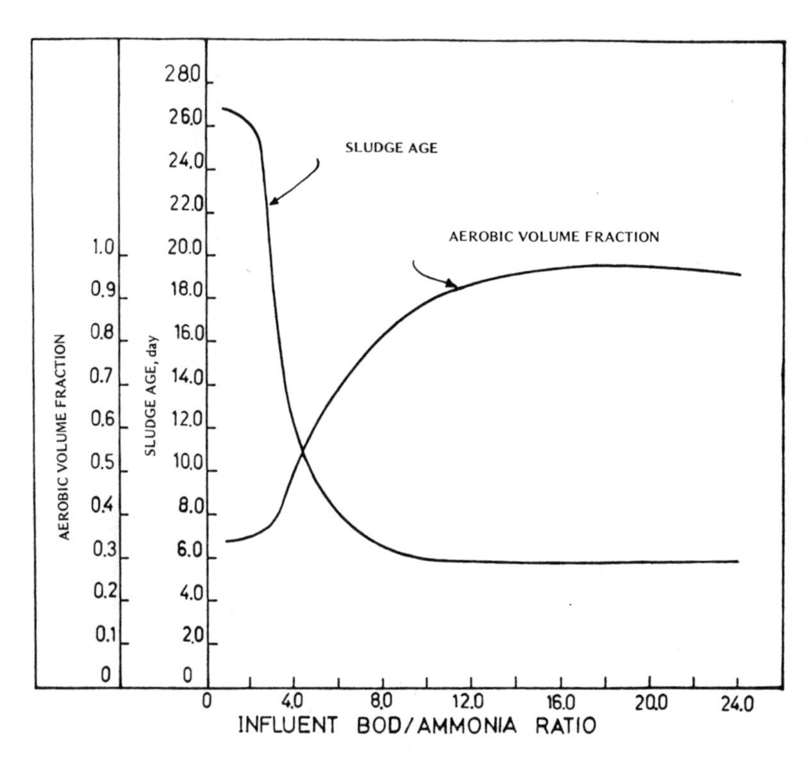

FIGURE 5. EFFECT OF INFLUENT COMPOSITION ON SLUDGE AGE AND AEROBIC VOLUME IN A RECYCLED SYSTEM

NUMERICAL EXAMPLE

Compare the two-sludge system with the single-sludge recycle system for the removal of BOD and nitrogen from wastewater with the following characteristics:

BOD = 500 mg/l
NH_3-N = 100 mg/l
NO_3-N = 0 mg/l
Temp. = 20°C

The effluent quality requirements are (soluble matter only):

BOD $\leq$ 50 mg/l
NH_3-N $\leq$ 5 mg/l
NO_3-N $\leq$ 15 mg/l

The following kinetic coefficients are used in this example all @ 20°C:

Nitrifier yield constant, a_N	0.15 $\frac{gVSS}{gNH_3-N}$
Heterotrophs yield constant (aerobic conditions), a	0.6 $\frac{gVSS}{gNH_3-N}$
Heterotrophs yield constant (in denitrification), a_{NT}	0.9 $\frac{gVSS}{gNO_3-N}$
BOD removal rate constant (aerobic), K_o	5.0 day^{-1}
Denitrification rate constant using sewage as carbon source, R_{DNM}	0.13 $\frac{gNO_3-N}{gVSS-day}$
Denitrification rate constant using methanol as carbon source, R_{DNM}	0.20 $\frac{gNO_2-N}{gVSS-day}$
Maximum nitrifiers growth rate, μ_N	0.45 day^{-1}

Endogenous decay rate, b:

heterotrophic microorganisms	0.02 gVSS/gVSS-day
autotrophic microorganisms	0.05 gVSS/gVSS-day

SINGLE-SLUDGE RECYCLED SYSTEM

The solution presented herein is an approximation derived by trial and error.

Nitrifiers fraction $f_N \cong \frac{(0.15)(100)}{(0.6)(500)+(0.15)(100)} = 0.048$

Assume $f_N = 0.05$

Assume $X_V = 2500$ mg/l

$X_{VN} = (2500)(0.05) = 125$ mg/l

$X_B = 2500 - 125 = 2375$ mg/l

Assume total detention time (based on through flow) is 12 hr (6 hr in the anaerobic stage and 6 hr in the aerobic stage)

$$\text{Approximate recycle ratio } R = \left[\frac{NH_3(\text{inf})}{NO_3(\text{eff})-NH_3(\text{eff})} - 1\right]$$

$$\frac{100}{15-5} - 1 = 9$$

Influent to denitrification stage:

$$BOD = \frac{(1)(500) + (9)(50)}{10} = 95 \text{ mg/l}$$

$$NO_3\text{-}N = \frac{(1)(0) + (9)(15)}{10} = 13.5 \text{ mg/l}$$

$$NH_3\text{-}N = \frac{(1)(100) + (9)(5)}{10} = 14.5 \text{ mg/l}$$

Removals in the first stage:

$$\Delta NO_3\text{-}N = \frac{(0.13)(2375)(6)}{(24)(10)} = 7.7 \text{ mg/l (denitrification)}$$

$\Delta BOD = 3.4\ \Delta NO_3\text{-}N = (3.4)(7.7) = 26.2$ mg/l(denitrification)

$\Delta NH_3\text{-}N = 0.035\ \Delta BOD = (0.035)(26.2) = 0.9$ mg/l (assimilation)

Influent to nitrification stage:

BOD = 95 - 26.2 = 68.8 mg/l

NH_3-N = 14.5 - 0.9 = 13.4 mg/1

NO_3-N = 13.5 - 7.7 = 5.8 mg/1

Ammonia Removal Rate $R_{NM} = \frac{0.45}{0.15} = 3.0 \frac{\text{g } NH_3\text{-N}}{\text{gVSS-day}}$

$$\Delta NH_3\text{-N} = \frac{(3.0)(125)(6)}{(24)(10)} = 9.4 \text{ mg/1 (nitrification)}$$

$$\text{Effluent BOD} = \frac{68.8}{1 + \frac{(5)(2375)(6)}{(500)(24)(10)}} = 43.2 \text{ mg/1}$$ (compared to 50 mg/1 as assumed)

Assimilated NH_3-N = (0.035)(68.8-43.2) = 0.9 mg/1

Effluent characteristics:

BOD = 43.2 mg/1 (assumed value = 50 mg/1)

NH_3-N = 13.4 - 9.4 - 0.9 = 3.1 mg/1 (assumed value = 5.0 mg/1)

NO_3-N = 5.8 + 9.4 = 14.2 mg/1 (assumed value = 15.0 mg/1)

Rate of Sludge Production, (expressed in mg VSS/1 of wastewater treated):

$$\frac{\Delta X_B}{Q} = (0.6)(68.8\text{-}43.2)(10) + (0.9)(13.5\text{-}5.8)(10) - (0.02)(2375)(0.5) = 153.6 + 69.3 - 23.8 = 199.1 \text{ mg/1}$$

$$\frac{\Delta X_{VN}}{Q} = (0.15)(9.4)(10) - (0.05)(125)(0.5) = 11.0 \text{ mg/1}$$

$$\frac{X_{VN}}{X_V} = \frac{11.0}{11.0+199} = 0.052 \text{ (assumed value = 0.05)}$$

TWO-SLUDGE SYSTEM

The first stage (nitrification) is designed on the basis of minimum sludge age, thus:

$$\theta_c \geq \frac{1}{\mu_N} = \frac{1}{0.45} = 2.2 \text{ day}$$

Hydraulic detention time:

$$t = \frac{(0.6)(500-50)(2.2)}{(2375)[1 + (0.02)(0.8)(2.2)]} = 0.24 \text{ day} = 5.8 \text{ hr}$$

(the biodegradable fraction of MLVSS is assumed to be 0.8)

BOD removal:
Assume first stage effluent BOD is 75 mg/l

$$S_r = \frac{(5)}{(500)}(75)(2375)(0.24) = 425 \text{ mg/l}$$

$$S_e = 500 - 425 = 75 \text{ mg/l}$$

$$\dot{\Delta}NH_3\text{-N} = \frac{(3.0)(125)(5.8)}{(24)} = 90.6 \text{ mg/l (nitrification)}$$

$$(\Delta NH_3\text{-N})\text{assimilation} = (0.035)(500-75) = 14.9 \text{ mg/l}$$

Effluent NH_3-N $\simeq$ 0 mg/l

$$\frac{\Delta X_B}{Q} = (0.6)(500-75) - (0.02)(2375)(0.24) = 255 - 11.4 = 243.6 \text{ mg/l}$$

$$\frac{\Delta X_{VN}}{Q} = (0.15)(90.6) - (0.05)(125)(0.24) = 13.6 - 1.5 = 12.1 \text{ mg/l}$$

$$\frac{X_{VN}}{X_V} = \frac{12.1}{243.6 + 12.1} = 0.048 \text{ (compared to assumed value of 0.05)}$$

Denitrification stage detention time:

$$t = \frac{90.6 - 15}{(2375)(0.20)} = 0.16 \text{ day} = 3.8 \text{ hr}$$

Methanol requirements:

$$M_e(OH) \simeq (3.0)(91-15) = 228 \text{ mg/l}$$

SUMMARY

The two alternative systems are depicted in Figure 6. As can be seen the recycled system has a total volume which is approximately 35 percent higher compared to the two-sludge system. It will also require more energy for pumping. On the other hand it contains only one clarifier and uses no chemicals.

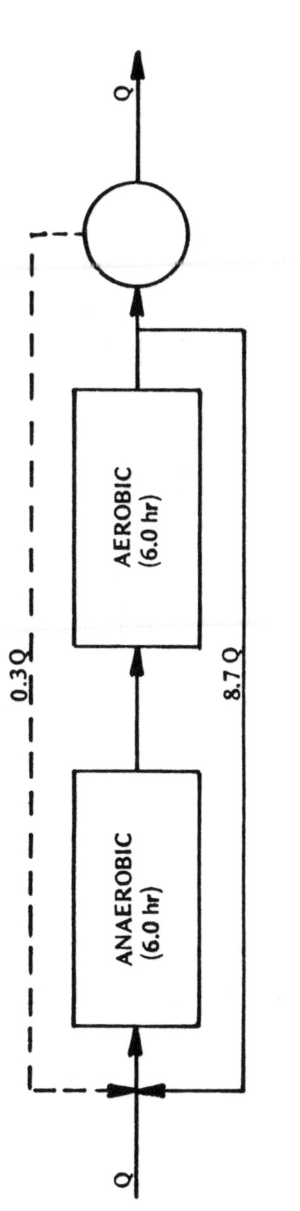

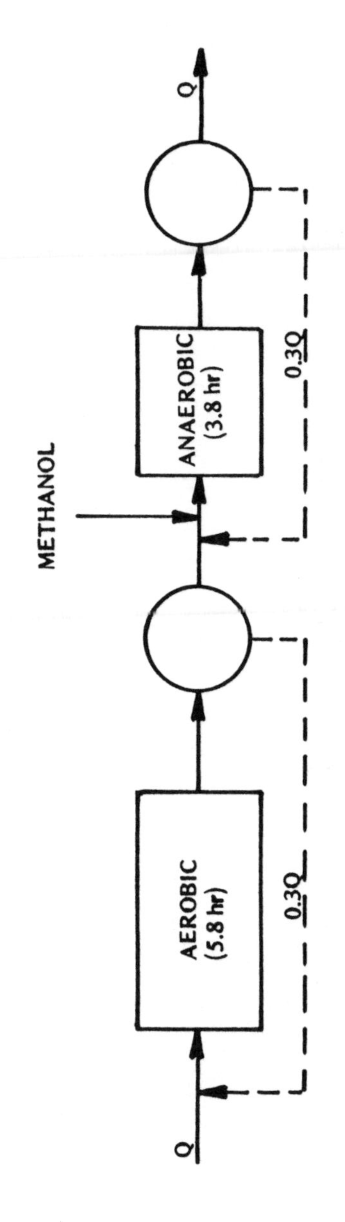

FIGURE 6. FLOW DIAGRAM OF ALTERNATIVE SYSTEMS IN NUMERICAL EXAMPLE

REFERENCES

1. U. S. Environmental Protection Agency, Process Design Manual for Nitrogen Control, USEPA, October, 1975.

2. Huang, C.S. and Hopson, N.E., "Nitrification Rate of Biological Processes," J. Env. Engng. Div., Am. Soc. Civ. Engrs. Vol. 100, EE2, pp. 409-422, April, 1976.

3. Wong Chong, G.M. and Loehr, R.C., "The Kinetics of Microbial Nitrification," Water Research, 9, 12, 1099, 1975.

4. Downing, A.L., Painter, H.A., and Knowels, G., "Nitrification in the Activated Sludge Process," J. Inst. Sew. Purif., 130, 1964.

5. Hockenbury, M.R. and Grady, C.P., Inhibition of Nitrification-Effects of Selected Organic Compounds, Journal of Water Pollution Control Fed. 49, 5, 768, May, 1977.

6. Anthonisen, A.C., Loehr, R.C., Prakasam, B.S. and Srinath, E.G., Inhibition of Nitrification by Ammonia and Nitrous Acid, Journal of Water Pollution Control Fed., 48, 5, May, 1976.

7. Christensen, M.H. and Harremoes, P., "Biological Denitrification in Water Treatment," Report 72-2, Department of Sanitary Engineering, Technical University of Denmark, 1972.

8. Wuhrmann, K., "Nitrogen Removal in Sewage Treatment Processes," Verh. Int. Verein. Limnol., Vol. 15, pp. 580-596, 1964.

9. Hunerberg, K. and Sarfert, F., "Experiments on the Elimination of Nitrogen from Berlin Sewage," Gas-u.Wasserfach, Wasser-Abwasser, Vol. 108, pp. 966-969 and 1197-1205, 1967.

10. Hamm, A., "Simultaneous Elimination of Nitrogen and Phosphorous," Z. Wasser-und Abwasser-Forschung, Vol. 3, pp. 102-107, 1970.

11. Schuster, G., "Laboratory Experiments on Removal of Nitrogen Compounds from Domestic Sewage," Fortschr. Wasserchem, Grenzgeb, Vol. 12, pp. 139-148, 1970.

12. Carlson, D., "Nitrate Removal from Activated Sludge Systems," Project A026 Report for OWRR, University of Washington, July, 1970.

13. Barnard, J.L., "Cut P and N Without Chemicals," Water and Wastes Engineering, Vol. 11, no. 7, pp. 33-36 and No. 8, pp. 41-49, 1974.

14. Matche, N.F., and Spatziemer, G., "Austrian Plant Knocks Out Nitrogen," Water and Wastewater Engineering, Vol. 12, (1), pp. 18, 1975.

15. Argaman, Y. and Miller, E., "Modeling the Recycled Systems For Biological Nitrification and Denitrification," Presented at the 50th Annual Conference WPCF, Philadelphia, Penn., October, 1977.

16. Wilson, T.D. and Newton, D., "Brewery Wastes as a Carbon Source for Denitrification at Tampa, Florida," Proceedings of the 28th Annual Industrial Waste Conference, Purdue University, Lafayette, Indiana, pp. 138-149, May, 1973.

17. Christenson, C.W., Rex, E.H., Webster, W.M. and Vigil, "Reduction of Nitrate-Nitrogen by Modified Activated Sludge," U.S. Atomic Energy Commission (TID-7517), pp. 264-267, 1956.

18. Climenhage, D.C., "Biological Denitrification of Nylon Intermediates Waste Water," Presented at the 2nd Canadian Chemical Engineering Conference, September, 1972.

19. Balakrishnan, S. and Eckenfelder, W.W., "Nitrogen Relationships in Biological Treatment Processes - III Denitrification in the Modified Activated Sludge Process," Water Research, Vol. 3, No. 3, pp. 177-188, 1969.

4. CARROUSEL ACTIVATED SLUDGE FOR BIOLOGICAL NITROGEN REMOVAL

H. David Stensel, Manager, Biological Systems, Eimco PMD, Envirotech Corporation, Salt Lake City, Utah

David R. Refling, Senior Research Engineer, Eimco PMD, Envirotech Corporation, Salt Lake City, Utah

J. Holland Scott, Process Engineer, Eimco PMD, Envirotech Corporation, Salt Lake City, Utah

INTRODUCTION

The Carrousel activated sludge system is described in terms of the system features, aerator application method, and method of nitrogen removal. Nitrogen removal levels and overall performance data experienced at sample Carrousel wastewater treatment plants are shown. The level of nitrogen removal is shown to be a function of the following design parameters: Activated sludge detention time and sludge age, mixed liquor suspended solids concentration, mixed liquor oxygen uptake rate and specific denitrification rate. Data is presented that shows nitrogen removals in the range of 40-70 percent for conventional Carrousel system designs. Achieving higher levels of nitrogen removal through increasing the system detention time is discussed. The design procedure used to determine nitrogen removal levels is described. Operational considerations and a description of an automated dissolved oxygen control system which can be used to maximize denitrification are presented.

Effluent quality requirements for wastewater treatment plants are, in many cases, being made more stringent. Some or all of the following treatment levels may be required:

- Effluent $BOD_5 \leq 10$ mg/l
- NH_3- ≤ 1.0 mg/l
- Nitrogen Removal
- Phosphorus Removal

These requirements usually result in increased plant capital and operating costs. The increased operating cost is normally due to increased power consumption and/or chemical costs associated with nitrogen and phosphorus removal.

The Carrousel activated sludge system can achieve such high quality treatment with savings in operating and capital costs. Effluent BOD_5 values of less than 10 mg/l, complete nitrification 40-70 percent nitrogen removal and a stable sludge for land disposal are common. Phosphorus removal can be accomplished by chemical addition to the Carrousel aeration tank.

The purpose of this paper is to describe the Carrousel activated sludge system and illustrate Carrousel designs to achieve nitrogen removal by biological denitrification.

CARROUSEL DESCRIPTION

The Carrousel activated sludge system was developed by Dwars, Heederik and Verhay, Ltd., Consulting Engineers in Holland in the late 1960's.[1] Their goal was to retain some of the advantages of the commonly used brush aerator oxidation ditch system, but to develop a more energy efficient and lower cost system using a deeper aeration basin. Since this initial work, over 128 Carrousel Activated Sludge Systems ranging from 0.5 mgd to 300 mgd in size have been put into operation.

A schematic of the Carrousel activated sludge system is shown in Figure 1. Fixed low speed surface aerators are applied in a special arrangement to aerate the activated sludge and promote sufficient velocities down the channels of greater than one foot per second to suspend the mixed liquor solids. The use of surface aerators provides more efficient oxygen transfer as well as efficient mixing in deeper channels as compared to conventional oxidation ditch systems. The aerators are positioned with one side of the impeller very close to the center partition to promote mixing. The channel velocity is dependent on a number of physical and dynamic factors such as impeller type and size, impeller RPM, aerator power draw, channel width, channel depth, number of aerators and tank volume. A mixing model has been developed to describe the channel velocity as a function of the above factors.

The Carrousel system combines some of the favorable aspects of both complete mixed and plug flow activated sludge treatment. The basis for complete mix operation is that at one foot per second channel velocity, the volume of liquid circulating through the channels may be between 30 and 50 times the influent flow. While an amount equal to the influent flow is continually displaced over the effluent weir, tremendous dilution is provided as the influent is combined with the mixed liquor recirculating through the channels. The liquid in the channels completes the circuit every 5 to 20 minutes depending on the channel length and design loading. This flow pattern prevents short circuiting and still offers the buffering features of a complete mix system.

The channel flow or plug flow operation provides operational benefits such as the ability to prevent short circuiting, denitrification at some distance from the aerators and improved settling and clarification in the final clarifier due to the gentle flocculation zone in the outer channel prior to the effluent withdrawal.

The Carrousel system results in a number of advantages in the utilization of aerators including efficient oxygen transfer rates and a very efficient means of mixing the activated sludge basin. With all of the aerators located at one end of the Carrousel aeration basin, a very high power intensity occurs in the aeration zone (roughly 4-8 hp/1000 $ft.^3$, as compared to .75-1.0 hp/1000 $ft.^3$ for conventional aeration). Figure 2 shows that the higher power intensity improves the aerator oxygen transfer efficiency.[2]

Only a portion of the aerator horsepower is

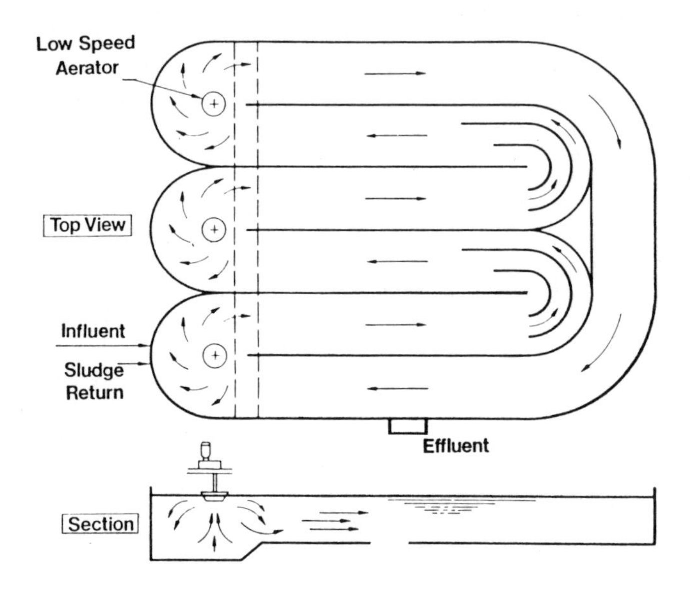

Figure 1.

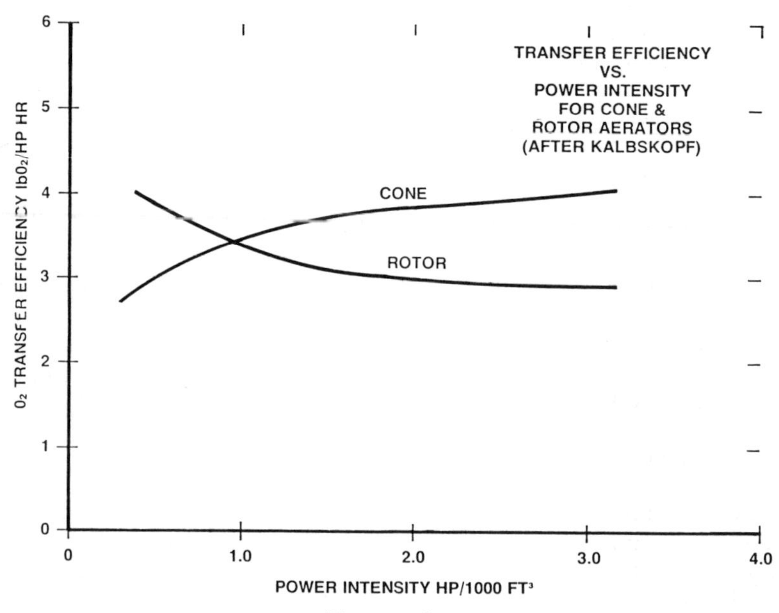

Figure 2.

generally used to generate channel flow, so that Carrousel aerators are normally designed on the basis of oxygenation requirements.[3] To produce a nitrified or extended aeration effluent, long activated sludge aeration times are normally required resulting in a mixing limited design for conventional activated sludge systems. The Carrousel system will not require the same level of mixing horsepower and thus significant energy savings are achieved in the production of a high quality effluent.

The aerator mixing capability can also result in a significant power savings during low loading periods. Figure 3 shows the ability to reduce aeration power requirements by 50 percent at Osterwoolde and still provide sufficient velocities for suspending the mixed liquor solids. This type of aerator layout makes it possible to use fewer aerators, reducing capital costs and maintenance requirements. Only two aerators need be used for up to a 5 mgd flow treatment unit. Carrousel aerators also afford an easy opportunity to cover the aeration zones for the prevention of misting and icing problems.

NITROGEN REMOVAL IN CARROUSEL

Due to the plug flow nature of the Carrousel flow scheme, the mixed liquor dissolved oxygen level is reduced to zero at some point in the channels. This promotes biological denitrification, as the bacteria utilize nitrate in place of oxygen for respiration. The amount of denitrification that occurs depends on the bacterial endogenous respiration rate of the mixed liquor and the length of the anaerobic channel section. As will be shown in the next section, these factors can be taken into account in design to provide desired levels of denitrification. In addition to nitrogen removal, denitrification results in the following benefits: (1) the use of nitrate to satisfy a portion of the oxygen demand. The net oxygen requirements can be reduced by 10 to 25% resulting in lower power requirements. (2) The restoration of some of the alkalinity destroyed during nitrification. About 3.6 mg alkalinity as $CaCO_3$ are returned per mg of nitrate nitrogen reduced to offset the 7.14 mg alkalinity destroyed per mg of ammonia nitrogen oxidized. This reduces or eliminates chemical additions for nitrification pH control. (3) Aids in minimizing the occurence of floating sludge by denitrification in the final clarifier. In addition to the low sludge activity due to a long sludge age

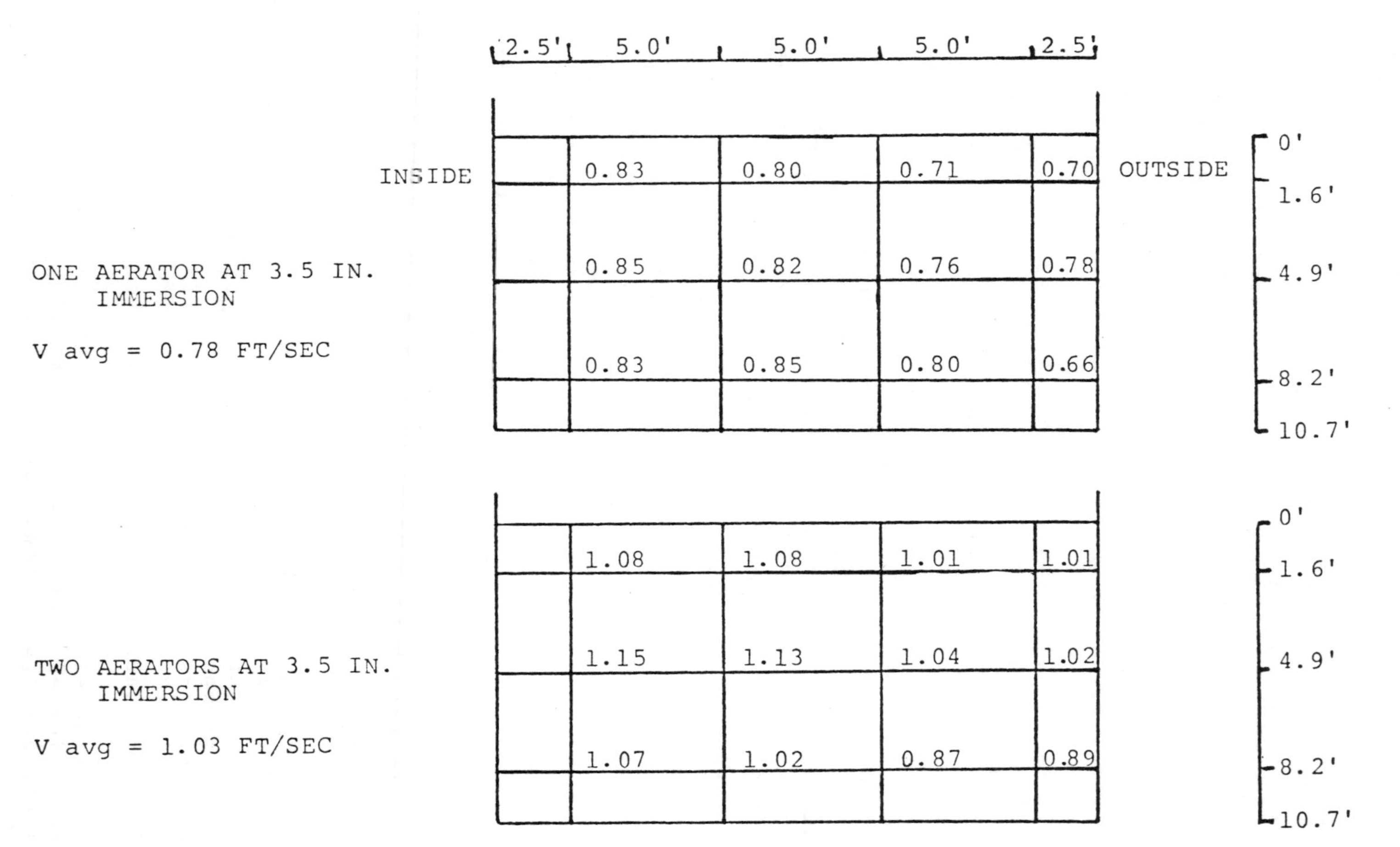

Figure 3. Channel velocity profiles at Oosterwolde.

design normally used in a Carrousel system, the low level of nitrates in the final clarifier essentially eliminates floating sludge problems.

Figure 4 shows the pathway followed by nitrogen in a Carrousel system designed for 70 percent nitrogen removal. Nitrification occurs in the portion of the Carrousel channels which are aerobic, and denitrification occurs in the length of channel where the dissolved oxygen is reduced to zero. The influent nitrogen is oxidized to nitrate or incorporated as organic nitrogen. In this generalized example, 60 percent of the nitrogen is removed by denitrification and 10 percent by sludge wasting. In order to control this system, automated dissolved oxygen control may be provided to adjust the aerators to insure that nitrification and denitrification volumes are adequate.

Table I shows nitrogen removal and performance results from some operating Carrousel installations in the Netherlands. Nitrogen removals ranged from 57-90 percent depending on the influent nitrogen concentrations and amount of denitrification that occurred. Assuming that the biological solids nitrogen synthesis requirements were 5 percent of the BOD_5, the nitrate production and percent denitrification could be estimated as follows:

Plant	Nitrate Produced (mg/l)	Percent Denitrification*
Winterswijk	40.5	47.5
Zutphen	40.4	49.6
Lichtenvoorde	74.4	28.6
N.T.F.	235.5	77.6

* Based on estimated nitrate reduced divided by total influent nitrogen.

These plants were not designed specifically for nitrogen removal, but the data shows significant levels of denitrification. The amount of denitrification depends on the design layout as well as operating methods and conditions.

DESIGN OF CARROUSEL FOR NITROGEN REMOVAL

The Carrousel system was originally designed to meet the strict effluent treatment in the Netherlands. An effluent tax is imposed, based on the quantity

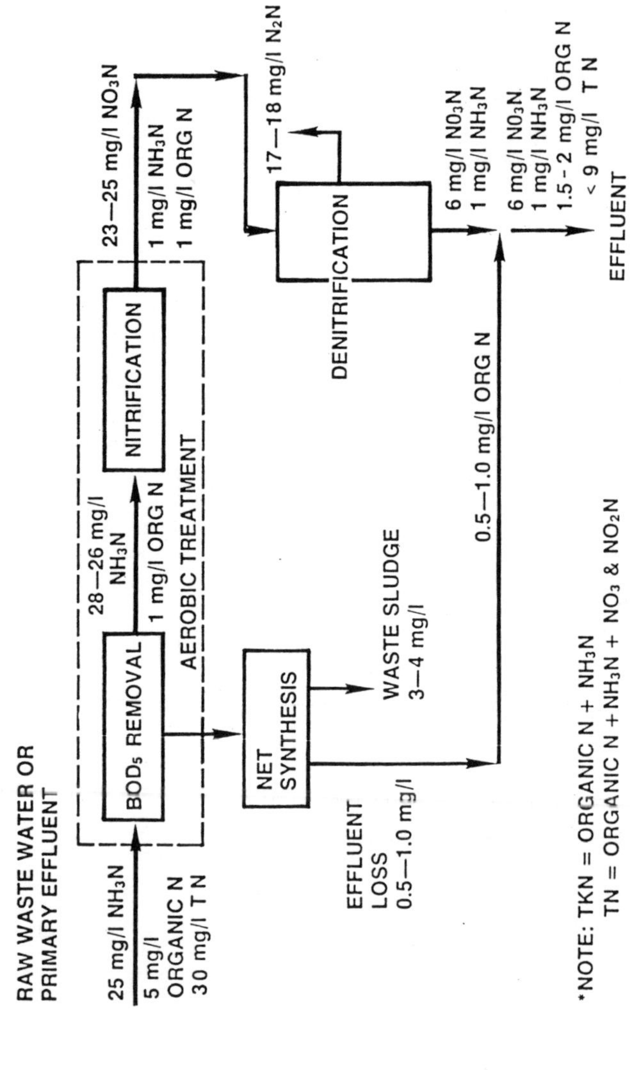
PATHWAY FOLLOWED BY TKN*
IN CARROUSEL
RAW WASTE WATER OR
PRIMARY EFFLUENT
25 mg/l NH_3N
5 mg/l
ORGANIC N
30 mg/l T N
BOD_5 REMOVAL
28—26 mg/l
NH_3N
1 mg/l ORG N
NITRIFICATION
AEROBIC TREATMENT
23—25 mg/l NO_3N
1 mg/l NH_3N
1 mg/l ORG N
NET
SYNTHESIS
WASTE SLUDGE
3—4 mg/l
EFFLUENT
LOSS
0.5—1.0 mg/l
0.5—1.0 mg/l ORG N
DENITRIFICATION
17—18 mg/l N_2N
6 mg/l NO_3N
1 mg/l NH_3N
6 mg/l NO_3N
1 mg/l NH_3N
1.5 - 2 mg/l ORG N
< 9 mg/l T N
EFFLUENT
*NOTE: TKN = ORGANIC N + NH_3N
TN = ORGANIC N + NH_3N + NO_3 & NO_2N

Figure 4.

TABLE I.

SUMMARY OF AVERAGE OPERATING RESULTS
FOR SOME CARROUSEL SYSTEMS

Winterswijk 16.5 MGD	Influent mg/1	Effluent mg/1	REMOVAL %
BOD_5	343	6.0	98.3
COD	714	53.0	92.6
Kjeldahl Nitrogen	59.5	3.9	93.4
Ammonia Nitrogen	-	1.9	-
Nitrate Nitrogen	-	8.3	-
Total Nitrogen	59.5	12.2	79.5
Zutphen - 6.25 MGD			
BOD_5	312	3.1	99.0
COD	738	38.0	94.9
Kjeldahl Nitrogen	56.5	3.3	94.2
Ammonia Nitrogen	-	0.5	-
Nitrate Nitrogen	56.5	12.4	-
Total Nitrogen	56.5	15.7	72.2
Lichtenvoorde - 3 MGD			
BOD_5	663	4.6	99.3
COD	1316	54.9	95.8
Kjeldahl Nitrogen	108.3	3.1	97.1
Ammonia Nitrogen	-	0.8	-
Nitrate Nitrogen	-	43.5	-
Total Nitrogen	108.3	46.6	57.0
N.T.F. Meat Rendering Bergum-3.8 MGD Equivalent			
BOD_5	750	5.0	99.3
COD	1050	-	-
Kjeldahl Nitrogen	277	6.9	97.5
Ammonia Nitrogen	-	4.0	-
Nitrate Nitrogen	-	20.4	-
Total Nitrogen	277	27.3	90.1

* Total Nitrogen = Kjeldahl Nitrogen plus Nitrate Nitrogen

of BOD and nitrogenous demand in the effluent. Because of this treatment requirement, Carrousel systems there have normally been designed and operated in the extended aeration mode. Nitrogen removal has not been a design objective. Different design procedures can be used to determine the nitrogen removal levels in a particular Carrousel application. The simplest approach is to design the Carrousel for a given sludge retention time (SRT) and then determine the resultant system anaerobic volume and denitrification capability. With the selection of the SRT, the quantity of oxygen and total aerator hp is determined. This power requirement will be reduced by the denitrification oxygen credit, once the level of denitrification is determined.

The selection of the Carrousel design SRT is normally based on the degree of treatment desired. If complete aerobic sludge digestion is desired, the Carrousel system SRT will be in excess of 20-30 days. If only nitrification is required, the SRT will be in the range of 10 to 20 days depending on the wastewater temperature. This is selected by applying a 2 to 1 safety factor to Downings temperature dependant equation for nitrification.[4] Denitrification requires an SRT in excess of that required for nitrification to provide anaerobic volume. Thus, the greater the level of denitrification required, the greater will be the required tank volume or SRT. The following equation is used to determine the necessary Carrousel Volume:[5]

$$V_A = \frac{(SRT)\ (Y_n)\ (\Delta\ BOD)}{MLSS\ \ (8.34)} \qquad (1)$$

Where:

V_A = Aeration volume, MG

BOD = BOD removed, lbs BOD/day

Y_n = net solids production coefficient, lbs TSS/lb BOD removed

MLSS = mixed liquor suspended solids concentration, mg/l

The MLSS concentration is normally selected to be 3500 mg/l to provide conservative solids floor loadings for secondary clarification. In Europe, these systems are commonly operated at MLSS levels of

4,000-7,000 mg/1.

One of the key design parameters in equation 1 is the net solids yield coefficient. A general set of curves have been developed for taking into account the synthesis yield coefficient, the endogneous respiration rate coefficient, and the effect of SRT and influent inert solids. Figure 5 summarizes a typical range of values for Y_n.

Inert solids are defined as solids entering the wastewater treatment system that are not biodegraded. These consist of inorganic solids and non-biodegradable volatile solids. The quantity of inert solids may range from 30 to 100 mg/1 depending on the degree of primary treatment. The higher the level of inert solids, the greater the net sludge yield and the greater the aeration tank volume necessary for a given sludge age.

The next step in the design is to determine the aerator horsepower requirements. This is done by determining the quantity of oxygen necessary for the level of treatment desired.

Figure 6 is an example of a curve used to select the amount of oxygen for BOD removal as a function of the design sludge age. In addition to this oxygen requirement, the amount of oxygen necessary for nitrification must be added. This is normally taken as 4.3 pounds of oxygen per pound of ammonia nitrogen oxidized. Once the oxygen requirements are determined, the aerator standard oxygen transfer efficiency value converted to a mixed liquor condition is used to determine the total horsepower for the Carrousel system.

Once the total horsepower is selected, the Carrousel basin layout can be evaluated. The number of aerators can be laid out in a variety of configurations as shown by Figure 7. Normally, the channel depth is in the range of 1.2 times the impeller diameter and the channel width is twice the depth. The hydraulic model developed by Dwars, Heederik, and Verhay, Ltd., is used to optimize the aeration basin and aeration design to provide maximum channel velocity. The design selected can be varied depending on a variety of operating and site location considerations.

For a given layout the aerobic volume is deter-

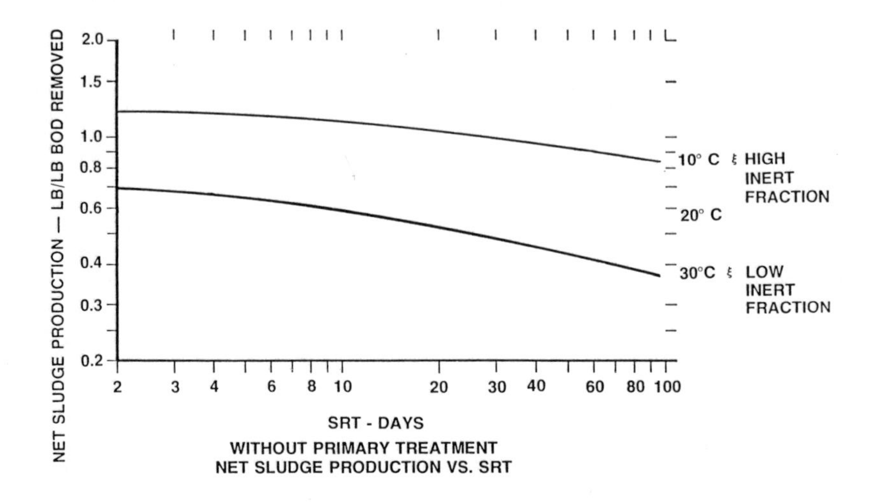

Figure 5.

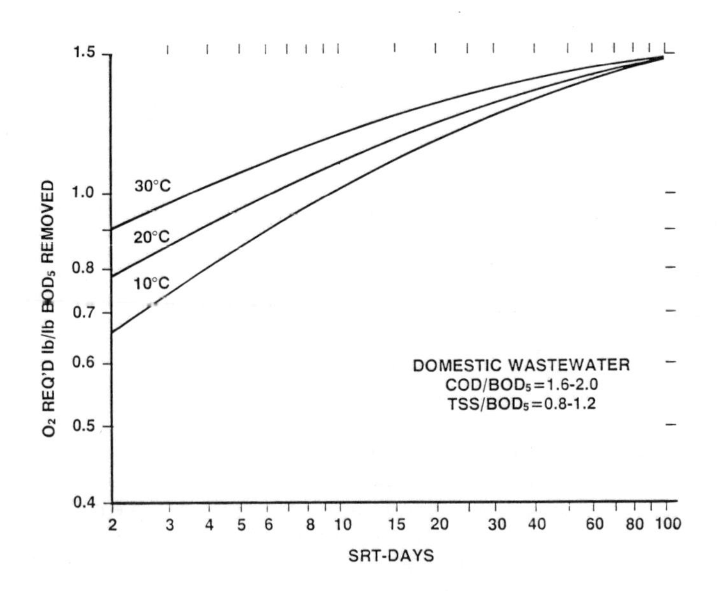

Figure 6.

mined and the location and extent of the denitrification volume is also determined. The aerobic volume is determined by dividing the dissolved oxygen concentration leaving each aerator by the system oxygen uptake rate and accounting for a channel velocity of one foot per second. The amount of nitrate reduced is then determined by the following equation:

$$N = V_D (DN_R) (MLSS)\ 8.34 \qquad (2)$$

Where:

N = Quantity of nitrate nitrogen reduced, lbs/day

V_D = Denitrification volume, MG

DN_R = Specific denitrification rate, lbs. NO_3-N per lb MLSS-day

The total quantity of oxygen required can then be decreased by a factor of 2.86 lb O_2 per lb. of nitrogen reduced. This will result in less total HP. In some cases, the aerator size numbers and layout must be re-evaluated to reflect this power reduction or it can be realized in operation by providing a lower power draw on the aerator by operating submergence selection.

NITROGEN REMOVAL DESIGN ILLUSTRATION

Table II shows influent wastewater characteristics assumed for a Carrousel design. The SRT of 25 days is selected to provide an aerobically digested sludge. The denitrification capability of this design will be illustrated. A denitrification rate of 0.025 is based on literature values for endogenous respiration denitrification.

Table III shows the total volume and horsepower determined by the design approach previously described. The location and size of the anaerobic zones are shown in Figure 8. There are two areas where the mixed liquor dissolved oxygen is depleted to zero to initiate denitrification. With the anaerobic volume available, at least 45 percent denitrification is possible. As Table IV shows the total nitrogen removal level is greater than 60 percent when cell synthesis is included. Two 60 hp aerators drawing 56.5 hp each at design is sufficient as the denitrification results in a 10 percent power savings.

CARROUSEL TANK CONFIGURATIONS

Figure 7.

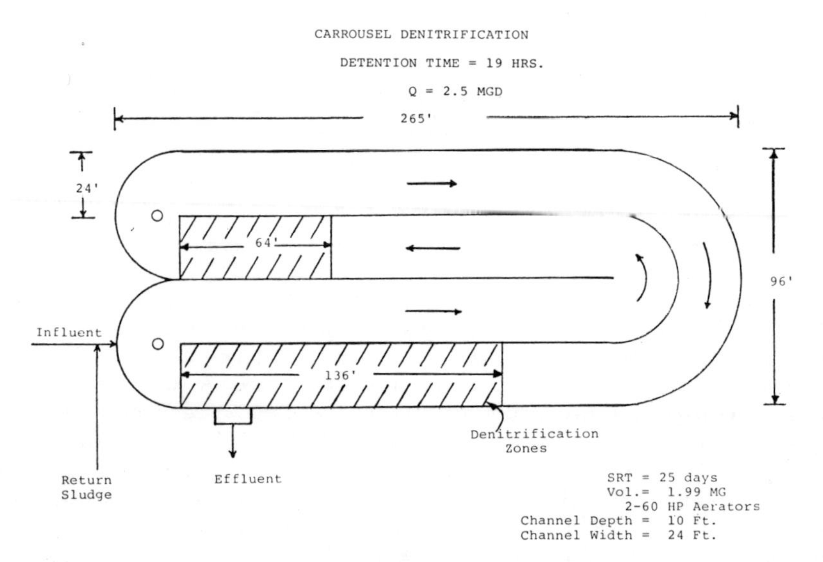

Figure 8.

Table V summarizes the design results if the SRT is increased to 30 days and the detention time increased to 24 hours. As Figure 9 shows the result is an increase in the denitrification zones, which results in 55% denitrification. The power savings is increased to 14%. Table VI shows that the total nitrogen removal can be in excess of 70 percent.

TABLE II.

DENITRIFICATION DESIGN EXAMPLE

INFLUENT WASTEWATER CHARACTERISTICS

FLOW	=	2.5 MGD
NO PRIMARY TREATMENT		
INERT SOLIDS	=	0.3 lb/lb BOD_5 REMOVED
BOD_5	=	200 mg/l
TKN	=	32 mg/l
TEMP. (MIN)	=	15^oC.
ASSUME: SRT	=	25 DAYS
MLSS	=	4000 mg/l
DENITRIFICATION RATE	=	0.025 mg NO_3-N mg MLSS-DAY

The amount of denitrification obtained in operation can actually be greater at times. This is due to the fact that the nitrification aerobic volume is greater than normally required. Thus, the aerator power draw can be decreased to increase the denitrification zone and save additional operating power.

An important observation on this type of system is that the necessary volume is provided for sludge stabilization, but a savings in operating power is possible while achieving a significant level of nitrogen removal. The degree of denitrification will

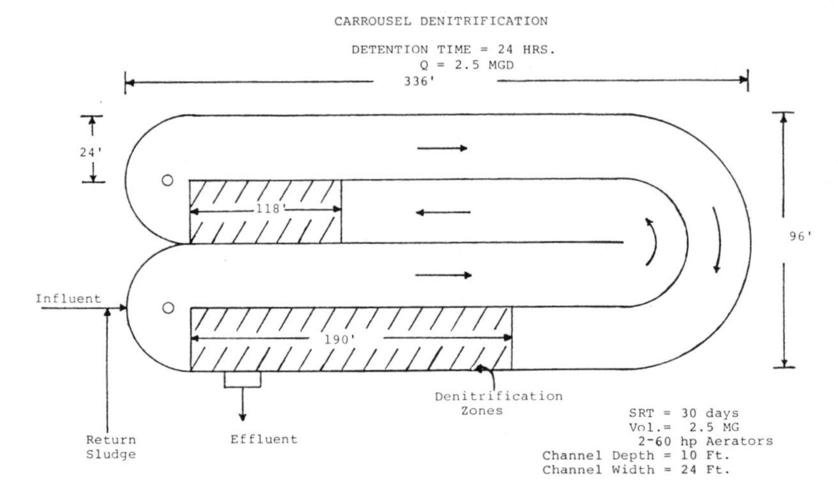

Figure 9.

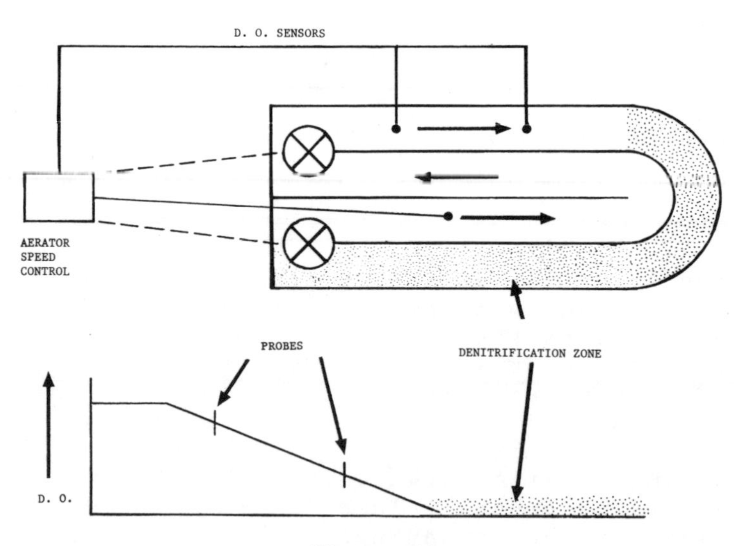

Figure 10.

TABLE III

DESIGN RESULTS

BASIC DESIGN		
TOTAL VOLUME	=	1.99 MG
DETENTION TIME	=	19 HRS.
OXYGEN REQUIRED	=	322 LB/HR.
POWER REQUIRED	=	126 HP
AERATORS	=	2-60 HP UNITS
TOTAL SLUDGE PRODUCED	=	2773 LB/DAY
DENITRIFICATION RESULTS		
ANAEROBIC VOLUME	=	0.34 MG
DENITRIFICATION	=	14.4 mg/l
OXYGEN CREDIT	=	36 LB/HR
FINAL HP REQUIRED	=	113 HP
NET POWER SAVINGS	=	13 HP
	=	10%

depend on the system detention time or SRT, the specific denitrification rate, the mixed liquor suspended solids level and oxygen uptake rate for the system.

DENITRIFICATION CONTROL

A major factor in maintaining a desired denitrification level is to assure that the anaerobic volume or point zero dissolved oxygen is met at all time. This has normally been done by using a single dissolved oxygen electrode at a point in the channel. The problem with this method is that it can not accurately control the denitrification zone since the

TABLE IV

DESIGN PERFORMANCE

SRT	=	25 DAYS
EFFLUENT PARAMETER		
BOD_5	$\leq$	10 mg/l
SS	$\leq$	20 mg/l
TKN	$\leq$	3 mg/l
NH_3-N	$\leq$	1 mg/l
NO_3-N	$\leq$	10 mg/l
TOTAL NITROGEN		
REMOVAL	$\geq$	60%
TOTAL DENITRIFICATION	$\geq$	45%

slope of the dissolved oxygen depletion determines this point, and one electrode cannot measure this slope. To solve this problem, multi dissolved oxygen probes are used to measure the dissolved oxygen level at two points in the channel as shown by Figure 10. This then determines the oxygen uptake rate since the velocity or time between the probes is known. The aerators can then be adjusted to raise or lower the initial dissolved oxygen level. This is done by varying submergence or aerator speeds.

SUMMARY AND CONCLUSIONS

The Carrousel system has been described and its nitrogen removal capabilities and design approach illustrated. Nitrogen removal levels of 40-70 percent can be accomplished in conventional Carrousel designs by denitrification and cell synthesis. These levels of nitrogen removal are achieved in systems designed for a high quality effluent (BOD_5 <10 mg/l, NH_3-N <1 mg/l) and sludge stabilization. No additional capital or operating costs are required to provide a significant level of nitrogen removal. In fact, operating costs are reduced due to aerator power

TABLE V

INCREASED DETENTION TIME

RESULTS

TOTAL VOLUME	=	2.5 MG
DETENTION TIME	=	24 HRS.
OXYGEN REQUIRED	=	327 LB/HR
POWER REQUIRED	=	128 HP
AERATORS	=	2-60 HP UNITS
SRT	=	30 DAYS

DENITRIFICATION RESULTS

ASSUMED DENITRIFICATION RATE	=	0.02 mg NO_3-N/mg MLSS/DAY
DENITRIFICATION	=	17.7 mg/l
	=	55%
OXYGEN CREDIT	=	44.3 LB/HR
FINAL HP REQUIRED	=	110 HP
NET POWER SAVINGS	=	14%

savings and to possible chemical savings for alkalinity control for nitrification.

The level of nitrogen removal is a function of the system detention time, mixed liquor solids level, specific denitrification rate and mixed liquor oxygen uptake rate. A variety of designs can be evaluated to achieve different levels of denitrification.

A two probe automated dissolved oxygen control

TABLE VI

DESIGN PERFORMANCE

SRT = 30 DAYS

EFFLUENT PARAMETER		
BOD_5	$\leq$	10 mg/l
SS	$\leq$	20 mg/l
TKN	$\leq$	3 mg/l
NH_3-N	$\leq$	1 mg/l
NO_3-N	$\leq$	6.7 mg/l
TOTAL NITROGEN REMOVAL	$\geq$	70%
DENITRIFICATION	$\geq$	55%

system can be employed to maintain necessary denitrification volumes and also maximize power savings.

REFERENCES

1. Koot, A.C.J., and Zeper, J., "Carrousel, a New Type of Aeration System with Low Organic Load", Water Research, 6, 1972, 401.

2. Kalbskopf, J.H., "Requirements of Aeration and Circulation Systems within the Scope of the Development of the Activated Sludge Process and Environmental Protection".

3. Jacobs, Allan, "A New Loop in Aeration Tank Design", paper presented at the New York Water Pollution Control Association Meeting, June, 1964.

4. Downing, et.al., "Nitrification in the Activated Sludge Process", Journal Proc. Inst., Sewage Purification-63 (2), 1964.

5. Stensel, H.D., and Shell, G.L., "Two Methods of Biological Treatment Design", Journal Water Pollution Control Federation, Volume 46, No. 2, p. 281, February 1974.

5. NUTRIENT REMOVAL IN THREE-STAGE PROCESSING

David W. Bouck, Dawkins & Associates, Inc.,
Consulting Civil and Environmental Engineers,
Orlando, Florida

ABSTRACT

Conventional wastewater treatment processes have little effect in the removal of nutrients, other than the incidental removal for biological cell synthesis during the relatively short treatment period. Specially designed treatment units utilizing biological, chemical and physical processes have been adapted to conventional treatment trains to predictably remove nutrients, such as phosphorus and nitrogen compounds.

One of the most effective of these process trains is a three-stage system utilizing biological processes to sequentially satisfy the carbonaceous BOD of the wastewater, convert ammonia nitrogen to nitrate form and subsequently reduce the soluble nitrate to nitrogen gas, which is then expelled to the atmosphere. In conjunction with these biological unit operations, chemical and physical processes are systematically promoting chemical coagulation and flocculation of soluble phosphorus compounds, for removal either separately or in combination with the biological waste sludge.

The system design approaches are widely varying, adaptable to many treatment process trains and offer extreme process flexibility to the plant operator. Of course, the operator engaged in advanced wastewater treatment technology contends with a greater number and complexity of unit operations. Process reliability is often increased by employing instrumentation and control systems designed to assist the operator or take direct control of particular functions.

INTRODUCTION

Historically, treated wastewaters have been discharged into a watercourse to utilize its dilution and assimilative capacities to maintain health standards and promote satisfactory environmental conditions. Over a number of years, the dilution capacity of many watercourses has reached saturation. At the same time, an increase in water consumption has required an ever-increasing volume and frequency of water re-use, with a resulting demand for increased levels of treatment.

Treatment and reuse of wastewater has become important, as the relatively fixed natural water supply is being overwhelmed by the demands of an expanding population. Even now, treated wastewater reuse ranges from agricultural and industrial process water, to coastal groundwater barriers to prevent salt water intrusion, to use as a direct source of potable water supply, with the degree of purification dictated by the specific use.[1]

Several states, including Florida, have placed more emphasis on wastewater treatment methods to curb pollution resulting from discharge of treated wastewater, especially in the state's interior fresh waters and sensitive coastal estuaries. This emphasis has resulted in water quality standards requiring advanced wastewater treatment (AWT) methods to reduce the discharge of conventional pollutants to a minimum and to also remove plant nutrients, such as nitrogen and phosphorus.

Often, conventional treatment methods cause minimal removals of plant nutrients, other than the incidental removal caused by biological cell synthesis during the relatively short treatment period. Biological processes have been developed which exploit the removal of nutrient compounds by cell synthesis, by establishing conditions which enhance the uptake and removal efficiency of naturally occurring organisms and plants.

Specially designed unit operations consisting of biological, physical and chemical processes have been devised that can predictably remove these nutrients. Typically, these treatment units have been adapted to conventional primary and secondary processing. The following is a review of unit operations designed to systematically convert and/or remove nutrients to acceptable AWT levels.

PHOSPHORUS REMOVAL

This section will deal with natural and enhanced phosphorus removal gained from the use of conventional primary and secondary processes and phosphorus removal realized through the use of chemicals as an adjunct to biological systems.

Typically, the majority of phosphorus compounds in wastewater are soluble and, therefore, are removed only sparingly by plain sedimentation. Secondary biological treatment removes phosphorus, normally organic compounds such as ortho-or polyphosphates, by biological uptake. However, there is surplus phosphorus available in relation to the quantities of carbon and nitrogen needed for cell synthesis. Consequently, typical secondary treatment results in 20 to 40 percent removal,[1] leaving an average effluent concentration of 7 ppm of phosphorus.[2]

The removal of nutrients beyond the biological synthesis requirements has been termed "luxury uptake." While many conventional treatment plants remove nutrients under some conditions to a surprising degree of efficiency, attempts to quantify the processes involved, so that they can be applied elsewhere, have produced limited success. Factors which are required to be present for the occurrence of increased phosphorus removal are a plug flow reactor basin configuration, a slightly alkaline pH, the presence of adequate dissolved oxygen, low CO_2 concentration, and no active nitrification.[3]

Phosphates and inorganic nitrogen can be removed from wastewater by algal synthesis. However, full scale growing and harvesting of algae to remove nutrients has not been feasible or practical for widespread application. Biological problems of proper balance of carbon to nitrogen to phosphorus ratio, adequate sunlight and temperature control, physical restrictions of the large pond area required, and costly mechanical harvesting techniques have prevented algal photosynthesis from being promoted as a practical means of nutrient removal.[1]

In conjunction with biological treatment processes, chemical precipitation, using aluminum and iron coagulants and lime compounds, is commonly used in phosphate removal by chemical and physical means. Although chemical coagulation reactions are complex and have advanced many theories, the primary action is believed to be a combining of orthophosphate with metal cations to effect colloidal charge neutralization. Polyphosphates and organic phosphorus compounds present in wastewater are most often removed by being adsorbed on floc particles for subsequent physical removal by sedimentation.

The coagulation of phosphates in raw wastewater also precipitates considerable organic matter in primary clarification. Where waste pickle liquor or lime are used, the primary clarifier is a common point of addition, and phosphorus removals of from 60 to 80 percent have been achieved.[1] Sodium aluminate has also been used in similar instances to reduce influent phosphorus and organic matter to subsequent treatment processes. Multiple point addition of sodium aluminate and alum in conjunction has been shown to provide effective phosphorus removal. A system utilizing these chemicals is presently proposed for a large regional AWT plant for the easterly Orlando area.

When preceding a biological treatment process, chemical addition to promote phosphorus removal can be effective if the pH is not raised above about 9.5 or below about 6.0, where decreased efficiency of secondary biological treatment often occurs. If the process scheme is physical-chemical treatment, excess lime is often added to increase phosphorus removals at high pH levels. Subsequent recarbonation can be used to lower the pH. Single and double-stage lime treatment and recarbonation are commonly followed by the physical process of filtration to remove nonsettleable solids.[2]

Several installations have shown that the addition of alum or ferric salts to activated sludge processes can provide effective phosphorus coagulation. The coagulating chemical is added directly to the aeration basin mixed liquor and the chemical floc is settled in the final clarifier with the biological floc. The BOD removal efficiency of the combination chemical-biological floc is not adversely affected, if care is taken not to depress the pH of the mixed liquor below normal ranges. Aluminate can be used with low alkalinity wastewaters, or where nitrification is required, to help maintain the pH.

Similarly, excess lime added to the primary clarifiers for phosphorus precipitation may often raise the pH to 9.5 and above, without adversely affecting the activated sludge process. Microorganisms in the aerated mixed liquor produce sufficient carbon dioxide to maintain a neutral pH in the process.

When chemicals are added for phosphorus precipitation, the quantity of sludge production can be expected to increase up to double the amount of that resulting from the biological process alone. The additional solids result from the chemical precipitant sludge formed, as well as the improvement in plant effluent suspended solids removal efficiency caused by the coagulant action of the chemicals.[4]

Single-stage chemical addition processes are often inadequate to remove phosphorus to the AWT levels of 1 ppm or less. Chemical precipitation can also be applied following secondary treatment processes to ensure consistent phosphorus removal above 90 percent.[4] This physical-chemical operation is similar to that for potable water processing including rapid mix, flocculation, sedimentation and filtration. It also offers the advantage of allowing separation of biological and chemical sludges which may initiate more economical waste sludge treatment and disposal techniques. The South Lake Tahoe facility found recalcining of lime feasible primarily for tertiary lime sludge, because of problems encountered with organic content in primary or combined lime sludges.[1] The method of tertiary removal of phosphorus utilizing alum and sedimentation is fairly common in Florida.

Coagulation applications have shown that alum requirements to precipitate phosphorus from wastewater are greater than stoichiometric equations indicate. One of the reasons is that natural alkalinity of the wastewater is consuming a quantity of the alum, before the pH of the wastewater is depressed sufficiently to allow formation of the chemical precipitate floc. The molar ratio of Aluminum to Phosphorus is 1 to 1, with the weight ratio of commercial alum to phosphorus at nearly 10 to 1. However, for phosphorus reductions to 90 percent and greater, the weight ratio required in actual application is often in the range of 20 to 1, meaning that to remove 10 ppm of phosphorus the alum dosage may be as high as 200 ppm.[4]

As mentioned, iron compounds are also commonly used to precipitate phosphorus. "Commercially available iron salts are ferric sulfate, ferric chloride, ferrous sulfate, and waste pickle liquor from the steel industry."[1] As is experienced with aluminum salts, a larger amount of iron is required in actual coagulation than the chemical reaction predicts. Lime addition is often used to raise the pH and aid the speed of reaction of the iron salts.

The action of lime as a phosphorus coagulant differs from the hydrolyzing reaction of metal coagulants. When added to water it raises the pH, reacts with carbonate alkalinity and combines with orthophosphate in the presence of the hydroxyl ion. If sufficient lime is added, precipitation softening continues as magnesium hydroxide is formed. Lime dosages of 200 to 300 ppm, sufficient to raise the pH to a range of 10, are required to remove the major fraction of phosphorus in wastewater.[1]

NITROGEN REMOVAL

Nitrogen in municipal wastewater results largely from human sanitary wastes, but can be extremely variable, expecially when kitchen garbage grinders and food processing industries add significant contributions. The nitrogen in raw sewage is almost totally in the forms of ammonia and organic nitrogen, with negligible nitrate. Total content in typical municipal wastewater approximates 10 pounds of nitrogen per capita per year. Conventional primary and secondary treatment normally removes 40 percent or less. The kinetics of nitrogen conversion in the treatment process has largely been covered in another chapter.

Environmental problems resulting from significant discharges of nitrogen include the decrease of dissolved oxygen in receiving water resulting from the oxidation of ammonia nitrogen to nitrate, the toxic effect of ammonia on waterborne wildlife, the effect of nitrogen as a plant nutrient in eutrophication of, especially, lakes and estuaries, and, of more recent concern, the limitation of nitrate nitrogen in drinking water sources.[1] Nitrogen removal processes have recently become an extremely important part of wastewater treatment technology.

It is nearly certain that nitrogen removal processes will be at the forefront in wastewater treatment systems because of increasing removal requirements of nitrogenous oxygen demand (NOD). Stoichiometrically, to accomplish nitrification, each pound of nitrogen requires approximately five pounds of oxygen, thus it can exhibit a great oxygen demand on a receiving watercourse. Nitrogen removal is being required in some states and in various localities in Florida, while it is under consideration in other states.

Biological nitrification will occur in most aerobic biological treatment processes when promoted by suitable operating and environmental conditions. Normally this occurs when a facility is hydraulically under capacity with resulting longer detention times and when warmer sewage temperatures are present. The problems of rising sludge during sedimentation of activated sludge or standard-rate trickling filter solids were often attributed to denitrification, with rising gas bubbles inhibiting the settling process. The common cure for the problem was to prevent the nitrification of the waste initially and to decrease the solids residence time in the clarifier unit.[5]

Where the treated wastewater is to be discharged to lakes or estuaries with little mass exchange and significant detention times, nitrogen removal requirements year-round have become the rule rather than the exception. Many plants are being designed and constructed to achieve complete nitrification during warmer months when oxidation rates are highest and when, at least in the Northern states, receiving streams have reduced flows and assimilative capacity.[6]

Nitrification facilities expected to operate year-round should be designed for complete conversion of ammonia to nitrate at the most adverse combination of influent ammonia and temperatures and with the proper pH range considered. Biological nitrification destroys alkalinity in the wastewater, with the result that the pH may be depressed sufficiently to inhibit nitrification. Unless sufficient alkalinity is present in the wastewater or chemicals are added to maintain the pH in the proper range, the process can be slowed to very ineffective levels.[5] The type of chemical system utilized for pH control and its degree of sensitivity will depend largely on the chemical balance of the wastewater and the variation of influent ammonia. To avoid problems with pH control during the nitrification process, sufficient alkalinity should be present to leave an alkalinity of 30-50 ppm in the effluent after nitrification is completed.[6]

Biological nitrification and denitrification in a two-stage system, preceded by secondary treatment, can achieve 90 percent inorganic nitrogen reduction and greater than 90 percent total nitrogen removals. This three-stage system shown in Figure 1 is considered necessary in most locations because wastewater temperatures below 65°F (18°C) may inhibit the biological growth rates below that required for adequate process efficiency. Typically, nitrification can be promoted in a conventional activated sludge system by increasing the solids retention time to at least 10 days. Because this can often lead to problems with the sedimentation portion of the process, due

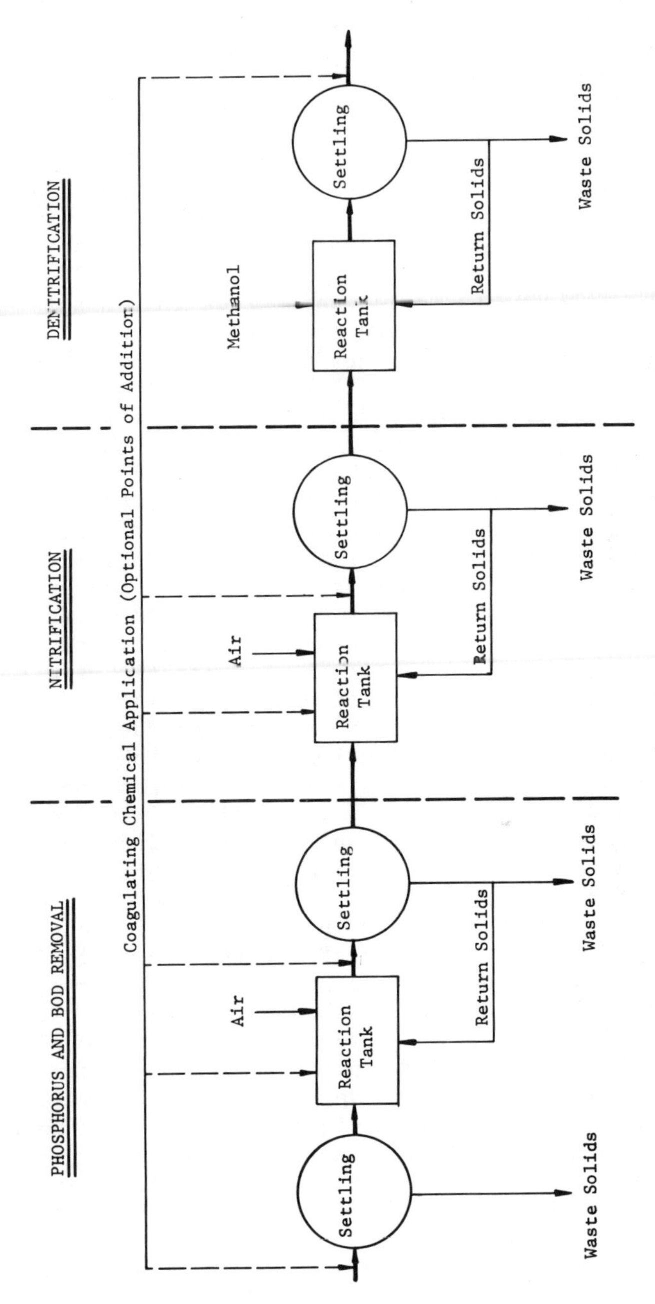

FIGURE 1. TYPICAL THREE-STAGE TREATMENT PROCESS FOR NUTRIENT REMOVAL

to formation of light pin floc or rising sludge, the process can often be more effective when divided to separate reactors. This method offers greater process flexibility; however, the plant operator must contend with a greater number of unit operations which can be a liability for any facility, especially for smaller communities.[3]

The first treatment stage, with or without chemical addition, is required to reduce the carbonaceous BOD to less than 50 ppm to allow nitrification to proceed in the second stage reactor. The third stage accomplishes biological denitrification utilizing bacteria, which feed on a supplemental carbon source and the oxygen present in nitrate nitrogen, to reduce the nitrate to nitrogen gas which is returned to the atmosphere.[6]

The system design applications are widely varying and adaptable to many treatment process trains. An advantage of the staged system is that it is adaptable as an additional treatment step to an existing secondary treatment facility, even if only required for seasonal treatment when water quality standards would otherwise be exceeded. Another advantage of biological nitrogen removal is that the nitrification process can be built to meet a current requirement for ammonia removal, and adequate performance can be achieved whether or not denitrification is immediately required.[2]

Further, should the requirement for nitrification be removed for any reason, the nitrification facilities can be converted easily to additional secondary process units at a minimal expenditure for treatment plant expansion. This was an important consideration in the design of the regional AWT plant for easterly Orlando, and the capability for process conversion was one of the design factors for process selection and operation.

The nitrate produced during nitrification can be reduced to nitrogen gas by a variety of facultative bacteria under anaerobic conditions. Because nitrified effluent contains little carbonaceous BOD for cell synthesis, an organic carbon source is required to act as a hydrogen donor and to supply carbon for biological synthesis of the denitrifying organisms. Numerous reduced organic chemicals have been successfully tested as a carbon source, including acetic acid, acetone, ethanol, methanol and sugar. Methanol has been the preferred substance in most applications, because it is one of the least expensive of the synthetic compounds.[1] It is easily miscible in water and promotes excellent growth of the organisms without the need for residual BOD in the process effluent. Methanol is nearly completely oxidized and consequently produces less biological sludge for disposal.

Because methanol is produced from natural gas, which is a resource of somewhat limited quantity, other by-products have been tested as oxidizable sources of carbonaceous matter.

Nitrogen deficient wastes such as brewery wastes and other wastewaters have been tried, but with limited success. The variations of the constituents added as carbon sources to the denitrification process and excess sludge generation have been common problems.[3]

Denitrification studies indicate that 3 to 4 pounds of methanol per pound of nitrate nitrogen are required to consume dissolved oxygen in the process stream and to sufficiently reduce the nitrate to nitrogen gas. Methanol demand for a typical domestic waste is about 60 ppm. The amount of methanol fed must be very closely controlled to ensure process stability and to avoid any excess dosage. Methanol is estimated to contribute about one-half of the costs of denitrification and excess feed often results in an undesirable residual BOD in the denitrification effluent.[5]

Reportedly, the most common denitrification system consists of a plug-flow basin, with a suspended-growth system agitated by low speed mixers to promote particle suspension without sufficient agitation of air estrainment. This is followed by a clarifier for solids separation and sludge removal.

Because of the area requirements of a third series of reaction tanks and clarifiers for suspended growth denitrification, alternative fixed-growth systems have been developed. These systems consist of static media upon which the denitrifying organisms contact the process stream. Typical fixed-growth systems consist of submerged, anaerobic filters (both single and dual media and with an up-flow or down-flow pattern), fluidized bed reactors which require considerable pumping energy to suspend and fluidize the supporting media, and packed towers with plastic or synthetic media of various configurations.

A recent improvement on the plastic packed tower is the application of a completely submerged plastic rotating biological contactor designed for denitrification service. It offers the advantages of a plug-flow tank design, excellent mixing and contact of the organisms with the wastewater, little short-circuiting or channelization (frequently encountered in packed towers) and with no need for a costly pumping system to provide the mixing energy of other systems. Although the fixed-growth system is reportedly less common than the suspended-growth systems, in truth, there are very few biological dentirification systems of either kind in full-scale operation. In the state of Florida, several fixed-growth systems are presently designed and/or are being constructed. At least one AWT facility with a fixed-growth denitrification process has several years of operating experience.

The process of denitrification results in the generation of carbon dioxide and nitrogen gas which must be purged from the wastewater. Because of the very gentle mixing of the denitrification tank designs, the process stream is commonly supersaturated with fine gas bubbles. Often, covered tanks have been employed to reduce surface aeration, because dissolved oxygen inhibits the reaction of the process and consumes the organic carbon source. Air-tight or walk-in covers should be avoided because of the nitrogen and carbon dioxide gases being expelled during the denitrification reaction.[5]

The denitrification system design for the large regional AWT plant for easterly Orlando will utilize a system of submerged rotating biological contactors in baffled tanks. The tanks will be covered by a layer of 100 millimeter high density polythelene spheres to reduce the tank water surface area by 92 percent, thus reducing the surface aeration and methanol consumption. The spheres also provide positive exit for the nitrogen gas being evolved and have the advantage of easy placement and/or replacement. They also require no secondary support, as in the case with a tank cover which accumulates rainfall and wind-borne debris. Laboratory scale studies with deoxygenated water show that a single layer of the spheres effectively prevents reoxygenation of the water even with considerable tank agitation.

When low solids concentration in the treatment plant effluent is required, tertiary filtration will be required following the treatment process discussed. The filter can be used to remove nonsettleable solids produced in upstream processes and is often aided by the addition of coagulating and/or flocculating chemicals to precipitate larger floc and aid the filtration of colloidal particles contributing to effluent turbidity. A subsequent chapter will discuss the mechanics and operation of the filtration process.

Several multistage systems for advanced wastewater treatment, including nutrient removal, are shown in Figure 2, 3 and 4. They utilize biological processes to sequentially satisfy the carbonaceous BOD of the wastewater, convert ammonia nitrogen to nitrate form and subsequently reduce the soluble nitrate to nitrogen gas, which is then expelled to the atmosphere. In conjunction with these biological unit operations, chemical and physical processes are systematically promoting chemical coagulation and flocculation of soluble phosphorus compounds, for removal either separately or in combination with the biological waste sludge. These systems have been designed to consistently reduce wastewater contaminents, including nutrients, to efficiencies exceeding 98 percent, with effluent concentrations below several parts per million.

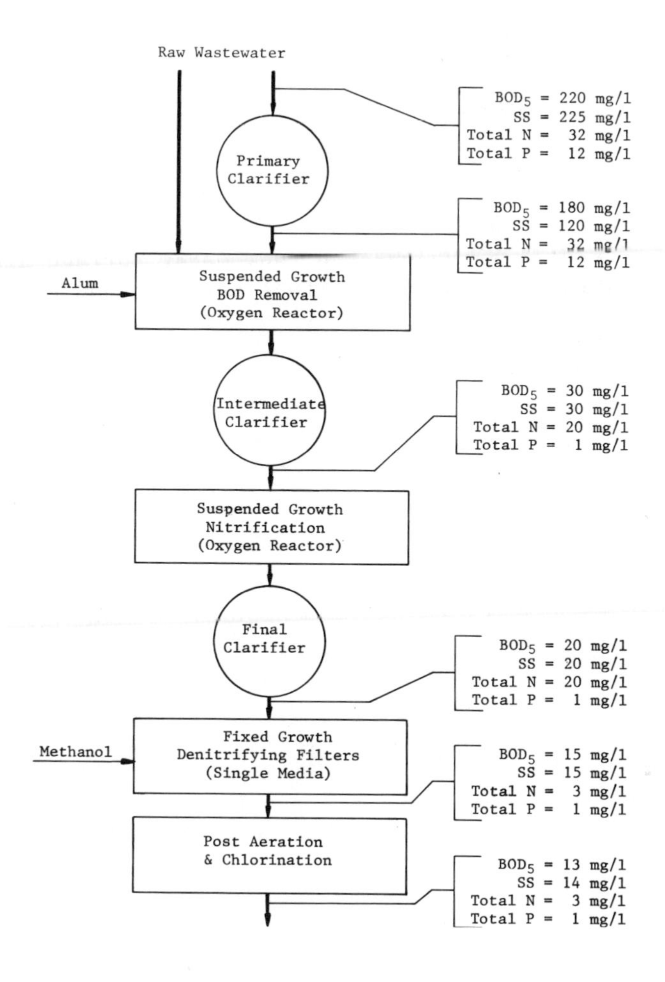

FIGURE 2. HOOKERS POINT PLANT--CITY OF TAMPA

The treatment process consists of partial primary clarification, followed by BOD reduction in a suspended-growth oxygen activated sludge reactor with alum addition for phosphorus removal, followed by intermediate clarification and solids recycle to BOD reactor, followed by nitrification in a suspended-growth oxygen activated sludge reactor, followed by final clarification and solids recycle to the nitrification reactor, followed by fixed-growth denitrification and filtration in a single media deep-bed filter with methanol addition for a carbon source, followed by post-aeration and chlorination for elevation of the effluent dissolved oxygen and disinfection.

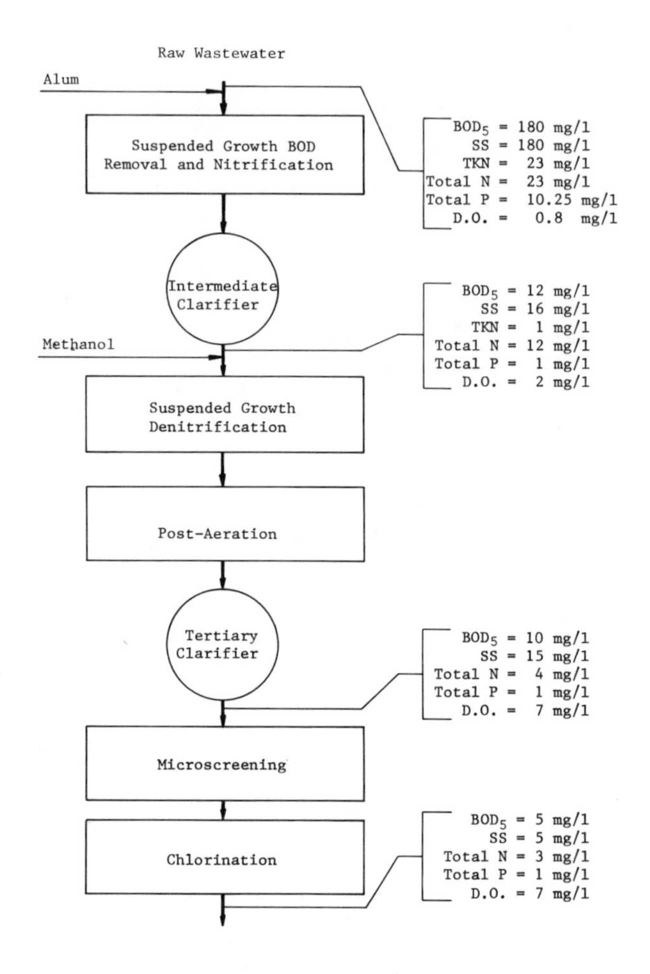

FIGURE 3. REGIONAL PLANT -- CITY OF ALTAMONTE SPRINGS

The treatment process consists of a suspended growth reactor for carbonaceous BOD reduction, with alum addition for phosphorus removal, followed by intermediate clarification for solids recycle to the nitrification reactor, followed by a suspended growth denitrification reactor, followed by post-aeration to elevate the effluent dissolved oxygen, followed by chemical flocculation, followed by tertiary clarification, followed by microscreening for residual solids removal, followed by chlorination for disinfection.

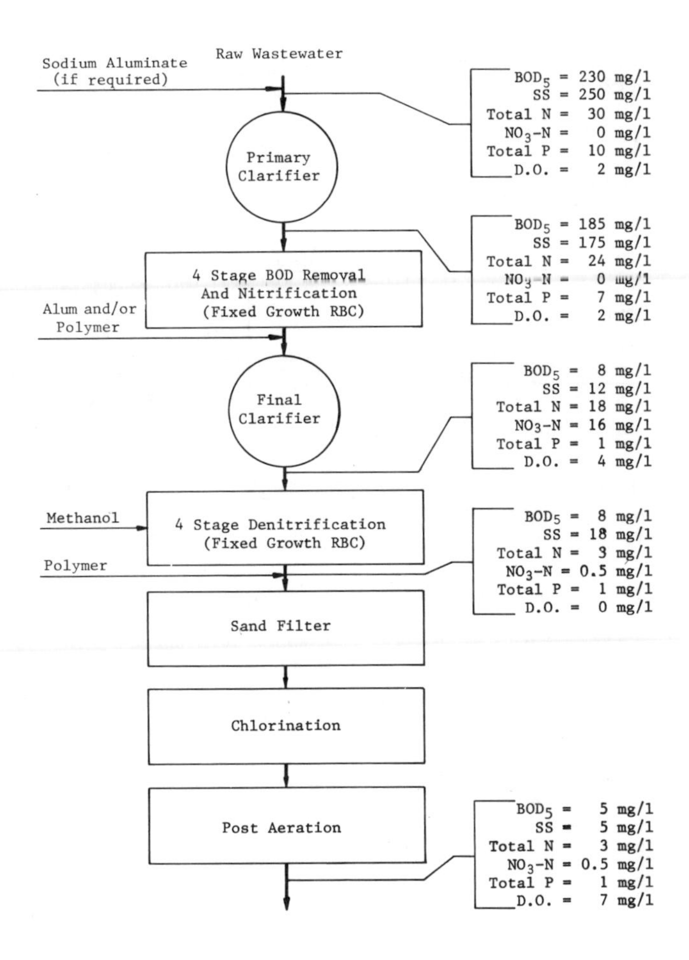

FIGURE 4. PROPOSED IRON BRIDGE ROAD PLANT -- CITY OF ORLANDO

The treatment process consists of primary clarification with sodium aluminate addition for organic loading control and partial phosphorus removal, followed by a four-stage fixed-growth system utilizing rotating biological contactors (RBC) and BOD reduction and nitrification, followed by final clarification with alum and/or polymer addition for phosphorus removal, followed by a four-stage system using a submerged RBC's for fixed-growth denitrification with methanol addition as a carbon source, followed by sand filtration with alum and/or polymer addition for residual solids removal, followed by cholorination for residual ammonia removal and disinfection, followed by post aeration to elevate the effluent dissolved oxygen.

REFERENCES

1. Hammer, M. J. Water and Waste-Water Technology (John Wiley & Sons, Inc., New York, 1975).

2. Metcalf and Eddy, Inc. Wastewater Engineering: Collection, Treatment, Disposal (McGraw-Hill Book Company, New York, 1972).

3. Wastewater Treatment Plant Design, Manual of Practice No. 8, Water Pollution Control Federation (Lancaster Press, Inc., Lancaster, Pa., 1977).

4. Process Design Manual for Phosphorus Removal. Prepared for the Office of Technology Transfer of the U.S. Environmental Protection Agency, 1976.

5. Process Design Manual for Nitrogen Control. Prepared for the Office of Technology Transfer of the U.S. Environmental Protection Agency, 1975.

6. Nitrification and Denitrification Facilities, Wastewater Treatment, Prepared for the Office of Technology Transfer of the U.S. Environmental Protection Agency, 1973.

6. THE BARDENPHO PROCESS

James L. Barnard. P.G.J. Meiring & Partners,
Pretoria, South Africa

This process for the removal of both nitrogen and phosphorus from sewage through the use of modifications of the activated sludge process was developed at the Laboratories of the National Institute for Water Research in Pretoria,[1] following good but inconsistent removal of nitrogen in systems such as the Huisman Orbal system and the Pasveer ditches, also presently used in the United States. Even though there is some resemblance to the system proposed by Ludzack and Ettinger,[2] the Bardenpho system was actually developed from an effort to apply the research of Balakrishnan and Eckenfelder[3] as described in an earlier publication.[1]

The process depicted in Figure 1 is a modification of an activated sludge process designed such that complete nitrification will take place at all operating temperatures. Nitrate-rich mixed liquor is recycled from the aeration basin to a basin ahead of the aeration basin where the incoming organic matter in raw or settled sewage can serve as hydrogen donors for the denitrification of the nitrates. When the mixed liquor is recycled at rates in excess of 5 times the average dry weather flow, removal of 80 to 85% of all nitrogen can be achieved. The effluent from the aeration basin that is not recycled is passed along to a second anoxic basin where the endogenous respiration of the cells will exert a demand for oxygen which can be satisfied by the nitrates remaining. The mixed liquor is then aerated before being passed through the clarifier. The underflow of the clarifier is returned to the first anoxic basin. The nitrifying bacteria remains in anoxic zones for short periods which do not seem to effect their

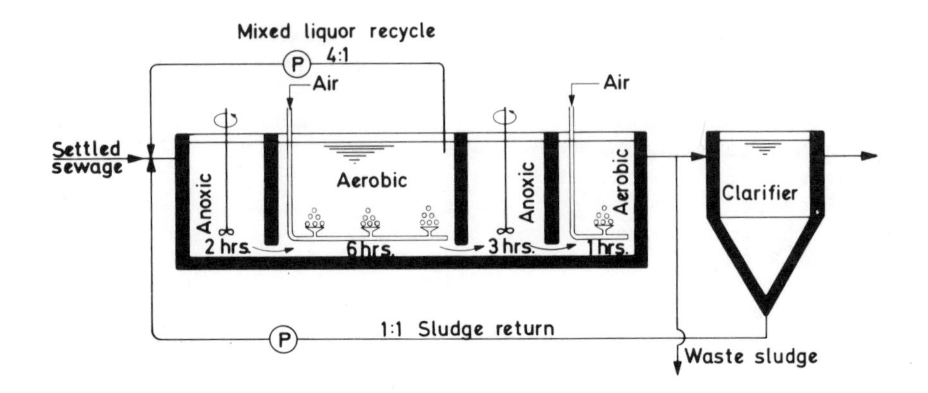

FIGURE 1. DIAGRAM OF BARDENPHO PROCESS

ability to convert ammonia to nitrate significantly. Nitrogen removals in excess of 90% is possible in this way without the addition of any chemicals. Typical results of a pilot plant operated in Pretoria is shown in Figure 2.

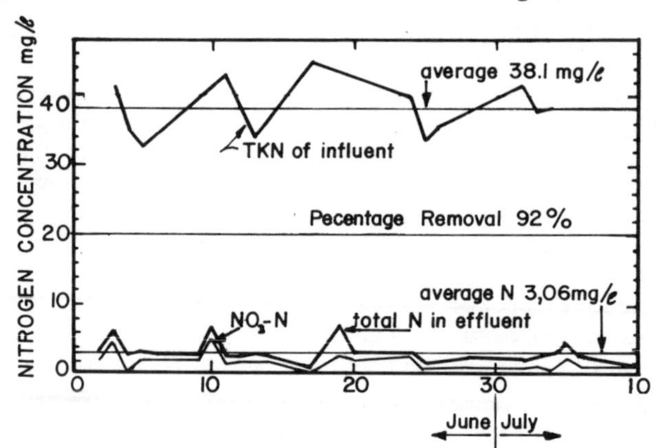

FIGURE 2. RESULTS OF 100 m^3/d PILOT PLANT DURING WINTER OPERATION

RATES OF DENITRIFICATION

Authors like McCarty,[4] Johnson[5] and Marais[6] discussed the rates of denitrification as affected by the organic compounds used for denitrification. McCarty presented rates for the use of various chemicals such as methanol, ethanol and acetate. A wide variety of rates have been quoted by other authors for the use of sewage as hydrogen donor for denitrification. The reason for this variation can probably be ascribed to the various combinations of organic matter in sewage. While trying to determine rate of denitrification in the pilot plant in Pretoria, mixed liquor was removed from the first anoxic basin and nitrates were added because of the low residual nitrate level. The rates of nitrogen removal was then determined by means of a nitrate probe. The curves in Figure 3 is typical of the rate determined in this way. In all cased there was a first rapid rate which lasted for about 5 to 6 minutes followed by the slower rate and this was

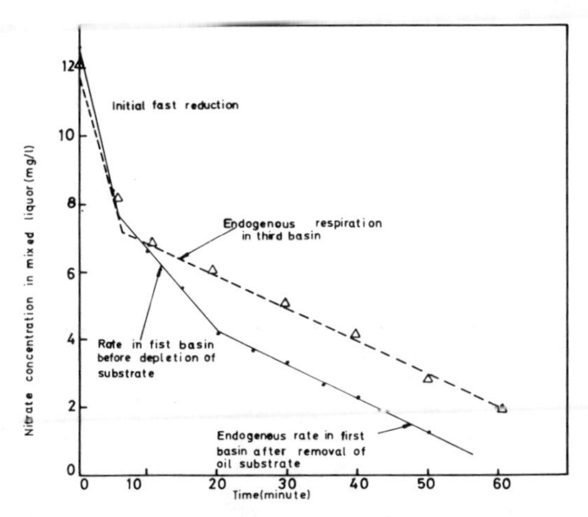

FIGURE 3. CONSECUTIVE FIRST-ORDER REACTION RATES FOR DENITRIFICATION USING SEWAGE AS HYDROGEN DONOR

followed by an even slower rate. It was immediately apparent that these rates were zero order reactions. In another set of experiments mixed liquor from the aeration basin was added in various proportions to equal volumes of the underflow from the clarifiers and the feed sewage. Again a first rapid rate of denitrification was observed, followed by a slower rate. Independently of the proportion of mixed liquor added to raw sewage the total mass of nitrogen removed per unit volume of raw sewage during the period that the rapid rate was operative remained the same irrespective of dilution. This indicated some substance existed in the raw sewage which would cause rapid denitrification but which was in limited supply. The initial rapid rate of denitrification was of the same order as that for acetate. It is well known that acetate is an intermediate formed during fermentation processes. It appears that during the passage of the sewage through the sewers and the primary tanks some acetate and other short chain fatty acid formation takes place. This means that the rate of denitrification can be accelerated significantly by feeding fermentation products into the denitrification zones. This is of special significance especially when the simultaneous removal of phosphorus is also desired.

Sutton[7] compared rate obtained through the range of operating temperatures by the author with that of other researchers as shown in Figure 4. The variation in the rates could perhaps be attributed to the different qualities of the raw feed. The mean unit denitrification rates as shown follows the dependency as given by Barnard,[8] ignoring the first rapid rates, as

$$D_N(T) = 3.6\ (1.09)^{T-20}\ \text{mg N/ g.MLSS/h}$$

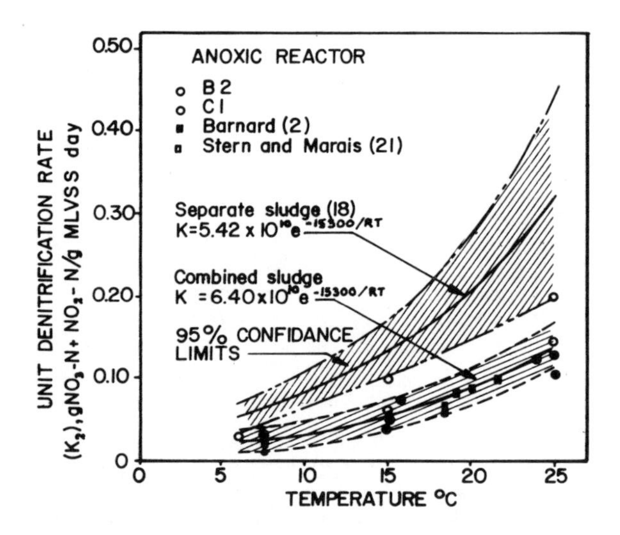

FIGURE 4. COMPARISON OF DENITRIFICATION RATES IN COMBINED AND SEPARATE SLUDGE SYSTEMS (AFTER SUTTON[7])

There seems to be a wide fluctuation in the rates obtained for endogenous respiration as obtained in the second anoxic basin. These rates are generally low and would not merit consideration except insofar as the effluent requirements demand that low total nitrates be obtained. The actual removal in the second anoxic basin may be as little as three to four mg/ℓ depending on the influent concentration and the sludge age. The values obtained by Barnard[8] seemed to be higher than those reported by other researchers.[6,7] These values were obtained by spiking of a sample of the mixed liquor removed from the second anoxic basin while under operation, which resulted in denitrification rates as shown in Figure 5.

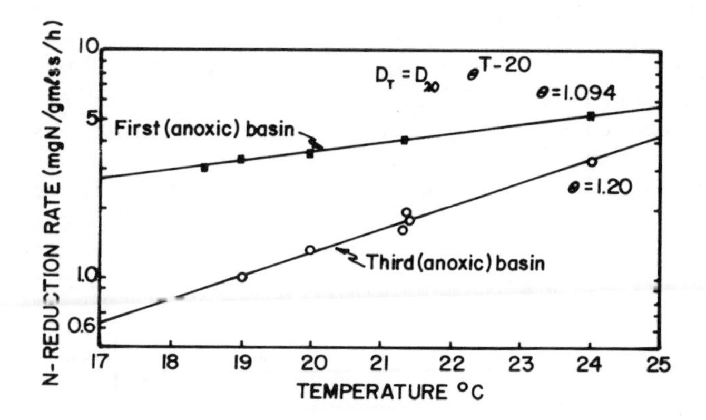

FIGURE 5. REMOVAL OF NITRATES IN THE SECONDARY ANOXIC BASIN

Since the pilot plant was removing between 92 and 95% of the nitrogen, the nitrate level in the second anoxic basin was low resulting in anaerobic rather than anoxic conditions. The fermentation products resulted again in a rapid rate of removal for a limited period, followed this time by a single rate. This last rate was reported as the actual denitrification rate which was found to be highly temperature dependent.

Due to basin limitations the pilot plant was operated in such a way that mixed liquor was passed on to the second anoxic zone shortly after emerging from the first anoxic zone as shown in Figure 5. While highly nitrified, it appears that much of the absorbed organic matter was not fully converted and was passed on to the second anoxic basin, causing higher than normal rate of endogenous denitrification. The mixed liquor was recycled from the far end of the aeration section thereby ensuring that the proper SRT for nitrification was maintained.

This would also expalin the extra-ordinarily high values for the endogenous denitrification rates obtained by Wuhrmann[9] who reported rates in excess of 14 mg NO_3-N/g.MLSS/h when using short retention times for nitrification. In actual plant design, it would not be possible to use these short retention times due to the safety factor involved for ensuring nitrification year round, taking into account the less than favourable conditions for nitrification that may occur such as inadequate dissolved oxygen (DO) at every point in the aeration basin, pH limitations, possible die-off of nitrifying organisms during the time spent in the anoxic and the fermentation zones required for phosphorus removal and the influence of toxic materials. The total oxidation of organic material in the aeration section could have been the reason for the lack of success of the application of the Wuhrmann process as discussed by Johnson.[5] In allowing sufficient aeration time to ensure good nitrification, while passing on the excess mixed liquor to the second anoxic basin shortly after the mixed liquor leaves the first anoxic basin would make maximum use of the absorbed organic matter for denitrification in the second anoxic basin. During the period that the pilot plant was operated as shown in Figure 6, the denitrification

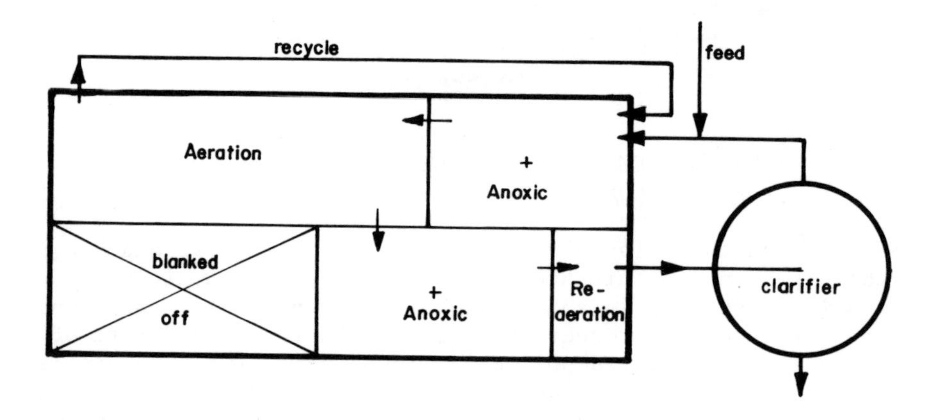

FIGURE 6. SHORTCIRCUITING OCCURING IN THE PILOT PLANT

rate in the second anoxic basin was so rapid that anaerobic conditions developed. This will be referred to again under the section for phosphorus removal. The operation was subsequently changed to allow the excess mixed liquor to be passed on to the second anoxic basin from the end of the aeration basin. Even though this did not have a markable influence on the total removal of nitrogen (possibly due to the basin being sufficiently large) the removal of phosphorus dropped markedly and consistently.

PHOSPHORUS REMOVAL IN THE BARDENPHO PROCESS

During tests with the Bardenpho system for the removal of nitrogen it was found that good phosphorus removal would also take place at times. There seemed to be a remarkable correlation between the presence of nitrates in the effluent and the inability of the plant to remove phosphorus. The graph in Figure 7 shown clearly this relationship between nitrates and phosphorus in the effluent. Extensive experimentation determined that for the removal of phosphorus to succeed, the bacteria must pass through a short period during which fermentation takes place. When nitrates are formed in an active sludge system and not removed, they will be present at all stages of the process. When oxygen is depleted the nitrates could sustain respiration albeit anaerobic and thereby prevent conditions of fermentation. It was therefore

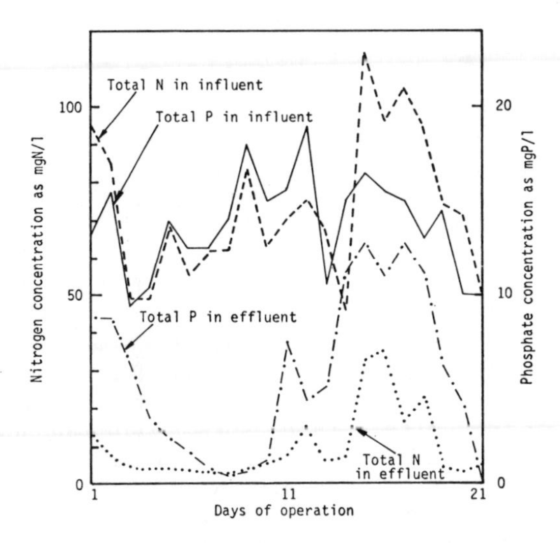

FIGURE 7. CORRELATION BETWEEN NITRATES AND PHOSPHORUS IN THE EFFLUENT

decided to add another small basin ahead of the first anoxic basin. The underflow from the clarifier will be returned to this first fermentation basin as well as some of the raw or settled sewage, while mixed liquor from the aeration basin will be recycled to the first anoxic basin as shown in Figure 8. In this way it can be assured that some fermentation will take place in this fermentation zone.

The application of this process in South Africa has lead to removal of phosphorus in the final effluent of activated sludge plants to a level between 0,2 and 0,8 mg/ℓ, together with removal of nitrogen of between 80 and 90%. Average effluent phosphorus values of 0,8 mg/ℓ as P with a standard deviation of 0,5 was possible in full-scale plants. In Johannesburg an extended aeration plant[10] treating about 100 000 m^3/d (26.4 mgd) of raw sewage was converted to this mode of operation by the switching off of aerators near the inlet zone in order to create denitrification and the necessary conditions for phosphorus removal. Average effluent concentrations of phosphorus were less than 1 mg/ℓ as P.

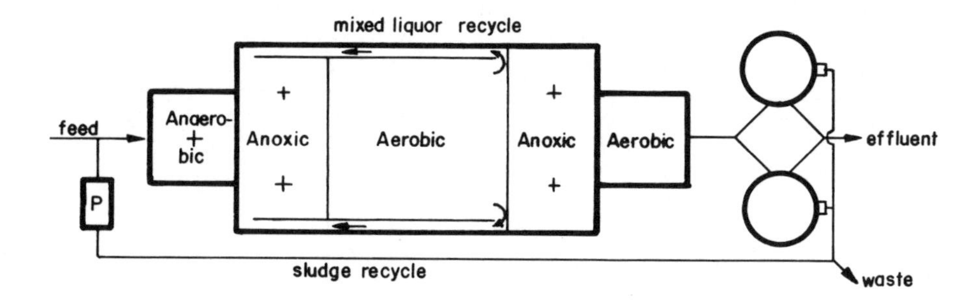

FIGURE 8. ADDITION OF FERMENTATION STAGE FOR PHOSPHORUS REMOVAL

POSSIBLE MECHANISM FOR BIOLOGICAL REMOVAL OF PHOSPHORUS

In South Africa we had to look at the biological removal to achieve our goals of reducing the eutrophication in receiving waters. We were aware that phosphorus was removed at certain plant in the United States such as at San Antonio[11] and Baltimore[12] and the Hyperion[13] Plant in Los Angeles. The pho-strip method of Levin[14] was also well publicised. The mechanism by which the phosphorus was removed remained unknown even after considerable study of the plants in the United States that removed phosphorus. Milbury[12] could operate a plant at Baltimore for more than 3 year to remove more than 90% phosphorus, but this achievement could not be repeated in other plants. All of the plants that removed phosphorus were of the plug type and in each case the sewage was fed only at one point. Milbury also noted that all plants experienced a release of phosphorus immediately after addition of the feed. Generally the feed would contain about 10 mg/ℓ as P and immediately after addition of the feed the dissolved phosphorus in the mixed liquor would reach values as high as 30 mg/ℓ while at the end of the aeration basin the concentration of phosphorus was about 0,5 mg/ℓ. Since all the plants were high rate plants little or no nitrification took place.

There are a number of theories relating to the ability of bacteria to take up phosphorus far in excess of their requirements of growth. Some theories held that bacteria that have been starved of phosphorus for a certain period will take up vast surpluses of phosphorus when exposed afterwards to liquid containing phosphorus under aeration. It was also determined that bacteria will store the surplus phosphates as volutinstaining poly-phosphates in the cells. The poly-phosphate can be discerned under the microscope when stained. Harold[15] and others describe the mechanism as being a shunt that occurs when no energy is available for the conversion of ADP to ATP as shown in this sketch in Figure 9.

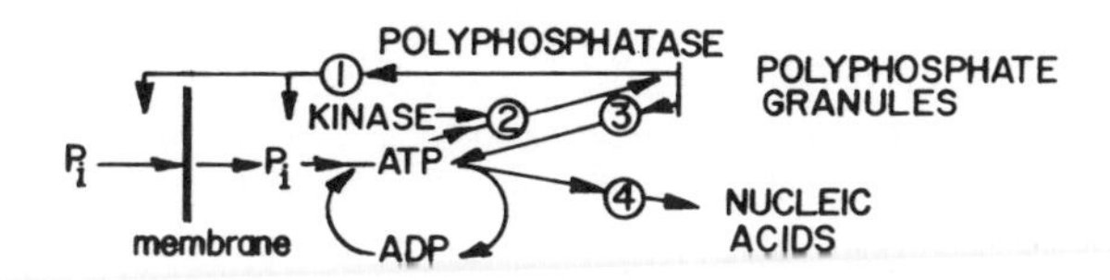

FIGURE 9. POSSIBLE PATHWAY FOR PHOSPHATE UPTAKE MECHANISM

Under these conditions the phosphate is stored as polyphosphate in the cell by the action of enzyme kinase. Under conditions of phosphate starvation with energy available, the cell will use phosphates from its stored pool. Under conditions of fermentation the cell will release phosphates from the surplus storage. Osborn[16] hypothesized that this release of phosphorus will create a disturbance in the cell's phosphorus balance which might trigger the mechanism for shunting the phosphates to storage under favourable conditions. At this point in time the mechanism is not fully understood but repeated test has shown that the mechanism can be triggered at will. It also appears that it is not necessary to remove phosphorus as is done in the Pho-strip method since all the phosphates released will again be taken up by the cell under the aerobic conditions. The phosphorus then leaves the system through the wasting of excess cells.

It becomes clear that in order to remove phosphorus, waste sludge must be dewatered fast without any phosphates being released to the liquid. This is best done by flotation thickening of the waste activated sludge before anaerobic conditions can develop and cause the release of phosphates. The waste activated sludge produced by plants having long SRT's can be flotation thickened easily without the addition of chemicals. The phosphorus contained in the cell would then still be available for algae growth or for the growth of food.

The removal of the phosphorus in those plants in the United States mentioned earlier could possibly have been caused by the lack of oxygen near the influent end of the aeration basin of these plug-flow plants. Nitrate formation in all these plants was depressed and with no nitrate available and with the possible lack of oxygen due to the high demand near the influent of the activated sludge plant and possibly also because of fermentation conditions existing already in the sewage reaching the plant, phosphate release took place in the inlet end of the aeration plant, resulting in the triggering of the mechanism for phosphate uptake.

In the pilot plant experiments referred to earlier, there was no fermentation zone and we had to rely on such conditions developing in the anoxic zones in order to trigger the mechanism of phosphate removal. This could naturally only take place when no nitrates were present in the effluent resulting in the good correlation between nitrate reduction and phosphorus removal. When the excess mixed liquor that was passed to the second anoxic zone contained some absorbed organic matter, the denitrification was complete and conditions were favourable for phosphate release. At such times up to 32 mg/ℓ phosphorus as P was measured in the filtrate of the mixed liquor from this zone. Rapid and immediate uptake of phosphorus followed in the re-aeration basin having a nominal retention time of one hour. When the plant was changed to allow the excess mixed liquor to enter the second anoxic zone from the end of the aeration basin, the phosphate removal dropped from 90% to less than 60% and little release of phosphorus took place.

PRACTICAL CONSIDERATIONS

In designing systems for the biological removal of nitrogen and phosphorus there are some practical considerations that must be taken into account such as SRT, temperature, sludge treatment and disposal, systems control.

Solids Retention Time

The solids retention time (SRT) is an important parameter to ensure that full nitrification will take place. Using the formulation of Downing,[17] the following relationship can be established

$$t_{xm} = 3.05\ (1.127)^{20 - T}$$

when t_{xm} is the minimum SRT for ensuring nitrification at $T^{o}C$. Since this condition will determine the SRT where the nitrifiers will be washed out of the system, one must apply a certain safety factor. In addition, the rate of nitrification will decrease with a decrease in the DO level below 1 mg/ℓ and with a decrease in the pH value below the optimum value of 8.3. The effect of the anoxic zones on the nitrifying organisms has not been established but Marais[18] feels that the aerobic zone must be larger than the combination of the anoxic and the fermentation zones. Due to lack of precise information, we have been using twice the SRT given above, using only the aerobic zone for this calculation. Thus for a lowest mixed liquor temperature of 15°C t_{xm} will be 5.5 days and design SRT 11 days. If the anoxic zones are say 40% of the total volume then the total SRT will be 18 days, not considering the sludge in the final clarifier.

The SRT also has a significant influence on the dewaterability of the waste sludge produced and the suspended solids of the effluent. Pitman[19] showed that the dewaterability of the waste sludge as represented by the capilliary suction time (CST) varies as shown in Figure 10, while the suspended solids of the effluent showed a similar variation. Due to the nitrogen and phosphorus associated with the suspended solids as well as the BOD, it is essential to design for low suspended solids in the effluent. Existing plant are usually operated as MLSS concentrations of between 4 000 and 5 000 mg/ℓ with SRT's varying between 15 and 20 days. This results in sludges that could be flotation thickened to solids concentrations of between 4 and 5% without the addition of chemicals, while producing effluents low in suspended solids and low in nutrients.

Sludge Treatment

The effect of the SRT on the treatment of the sludge leads into this section. For smaller plants, the most economical design of the system consists of the extended aeration application, having an SRT sufficiently long to ensure a stable sludge that could be dewatered on sludge drying beds or irrigated on land. Under these conditions, sludge is withdrawn directly from the aeration basin on a constant basis in such a way that the SRT is determined hydraulically. For very small plants, the mixed liquor can be discharged directly on to the beds with some decantation on the beds. The waste mixed liquor could also be passed to a small clarifier for some thickening of the sludge. For larger plants, the waste sludge is passed directly to a flotation unit for thickening to between 4 and 5% without chemicals, the underflow containing on the average no more than 50 mg/ℓ of suspended solids. The present policy for larger plants demands that the sludge be returned to land where possible and it seems to be most expeditious to discharge the thickened sludge to land. During wet seasons, filter belt presses could serve as a standby unit for dewatering. This would require the addition of about 2.5 kg of polyelectrolyte per ton of dry solids (5 lb/ton).

The process could be used with or without primary sedimentation and sludge digestion especially where the latter is required for reducing the energy demand of the plant. In such instances the two sludge streams are not mixed but treated separately.

If during any stage of the treatment of the secondary sludge containing high concentrations of phosphate, the sludge is allowed to undergo anaerobic conditions, phosphates will be released to the liquid stage which may not be returned to the biological plant without treatment for the precipitation of the phosphorus. Dissolved air flotation followed by belt pressing will not result in the release of phosphorus, while the underdrains of drying beds would need some treatment or be irrigated.

When the SRT of the plant is considered to be too short for producing a stabilized sludge, the waste mixed liquor could be further stabilized by the use of aerobic digestion. Hemstead and Marais[20] showed that applying the same principles of nitrification/denitrification and phosphorus removal at this stage will result in a negligible total release of phosphorus to the liquid. Stabilization of sludge from a plant removing phosphorus resulted in an overall release to the liquid phase of 15 mg/ℓ representing less than 12% of the phosphorus removed.

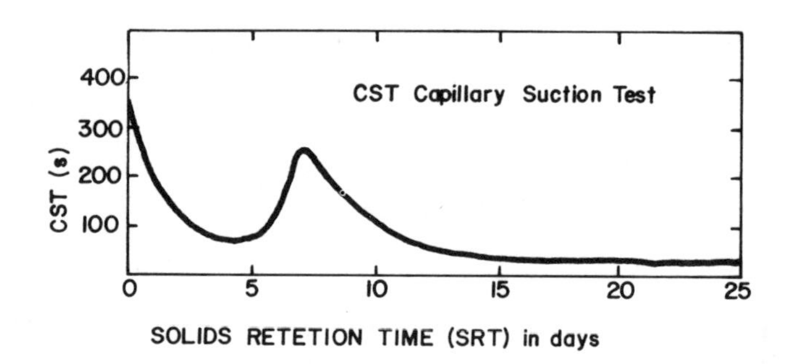

FIGURE 10. THE EFFECT OF SRT ON THE CST OF THE WASTE SLUDGE

Temperatures

In addition to the effect of the temperature on the minimum requirements for nitrification, the denitrification rates are temperature dependent and this must be reflected in the design as discussed earlier.

Hydrogen ion concentration

Nitrification is severely affected by a drop in the pH value. Also in soft water areas, the nitrification itself will destroy the alkalinity of the water. Generally, the formation of 1 mg/ℓ of NO_3-N will neutralize 7 mg/ℓ of alkalinity. The alkalinity will partially be restored by the destruction of the nitrates, but due to the similar pH effect on the denitrification rates, the designer must be careful to ensure that the pH value remains above 7 and preferably above 7.5. Recent experiments in Japan with industrial effluents having high concentrations of ammonia resulted in an overall reduction of the pH value when applying the Bardenpho process with resultant incomplete nitrification and denitrification. The pH was restored to 7.5 by the addition of lime which resulted in more than 95% total removal of nitrogen with more than 90% removal of phosphorus. In the regions around Cape Town, all extended aeration plants are designed with some form of denitrification to prevent drastic drops in the pH value when nitrification takes place.

The process could also be used directly after lime treatment without the need for neutralization. Laboratory scale experiments showed that lime treated raw sewage could serve as feed for the process resulting in extremely good removals of nitrogen producing a crystal clear effluent low in nutrients and colour.

Controls

For the proper functioning of the process it is essential that the plant be properly controlled. The control of the SRT was discussed earlier. In addition, the operator must check on the rate of recycle of the mixed liquor to the first anoxic zone and control the air supply to the main aeration basin. DO control is recommended also for saving energy.

The high rates of recycling of the mixed liquor requires that no unnecessary headloss be tolerated. In existing plants, typical headloss that needs to be overcome is of the order of 25 mm (one inch). Special pumps are required but the energy used is small even though the pumps are quite inefficient. Axial flow propellor pumps serve the purpose well. Dutch horizontal screw pumps could also be used for the purpose. Normal metering of the flow will not be possible and the recycle flow must be estimated from channel velocities. Accurate measurement is not essential. Means of adjusting the delivery of the pumps must be provided in order to optimize the operation.

Due to the influence of the nitrates on the plant's ability to remove phosphorus, it is essential that the nitrates be kept below a certain minimum level. In South Africa, the interest in the plant has moved from high nitrogen removal to phosphate removal. The degree of denitrification required to ensure phosphate removal would then be of interest. Generally some land treatment is incorporated or the effluents are passed through reed beds as is the case in Johannesburg. The remaining nitrates in the effluent is removed in a short section of the reed beds. Within 300 m from the plant there are only traces of nitrogen remaining in the effluent. However, experience showed that the nitrates in the effluent must be less than 5 mg/ℓ to prevent a complete breakdown of phosphorus removal.

The operator must also ensure that sufficient dissolved oxygen is passed on to the clarifiers to prevent anaerobic conditions which could result in the release of phosphorus to the liquid stage.

The fermentation zone must be stirred gently without the introduction of air. While the raw feed is essential for creating the necessary conditions in this zone, storm or other high flows will disturb the necessary conditions and must be by-passed.

In general the control of the process is fairly simple as can be seen from the number of plants in South Africa that are being manipulated for the removal of phosphorus while they were not designed for the purpose.

PRACTICAL APPLICATIONS

Shortly after the development of the process on laboratory scale units, the author operated a 100 m^3 per day (26 000 gallons per day) pilot plant to study the feasibility of the process. The main interest at this time was removal of nitrogen and while extremely good results were obtained during prolonged periods, as can be seen in Figure 2 it was not possible to run year long reliability tests due to luck of funds and lack of supervision of the plant. Average total nitrogen levels of less than 3.0 mg/ℓ were observed for extended periods. Phosphorus removal during this time was sporadic since the mechanism was not understood and studies were in progress to determine optimum conditions for phosphorus removal. Since completion of this study a better understanding of the phosphorus removal mechanisms lead to more emphasis being placed on the removal of phosphorus while the interest in nitrogen removal was confined only to its influence on phosphorus removal. The reason for this development was a discovery that nitrogen could easily be removed by growth in the rivers in the most affected areas.

The pilot plant has since been operated for a full year, this time with proper supervision especially to determine the reliability of removing phosphorus. The results will become available at the forthcoming IAWPR conference in Stockholm, Sweden, and I was informed that the average phosphorus level in the effluent of the plant was below 1 mg/ℓ. Average total nitrogen values varied between 3 and 5.5 mg/ℓ. There was a virtual complete corrolation between nitrates in the effluent and phosphate concentrations in the effluent. All indications where that better DO control as opposed to the system that was used, would have resulted in better overall results.

A number of large scale plants have been designed on the basis of the Bardenpho process, the largest of these the Goudkoppies treatment plant of the city of Johannesburg is being commissioned at this stage. In the interim period a number of plants have been manipulated in order to reduce the phosphorus content of the effluent. Some of these plant will be briefly described.

MODIFIED EXISTING PLANTS

Alexandra plant - Johannesburg

This plant was built to relieve the main sewer to the Northern works of Johannesburg temporarily. The sewer has been completed in the meantime but due to overloading of the Northern works, the relief works are now treating 45 Mℓ/d (12 MGD). The layout of the plant is shown in Figure 11. In the first basin 14 surface aerators operate without division walls between them. By turning off 4 aerators near the influent zone some recycling of mixed liquor was obtained due to the action of the other aerators, leading to an overall removal of 85% of the nitrogen entering the plant. Table 1

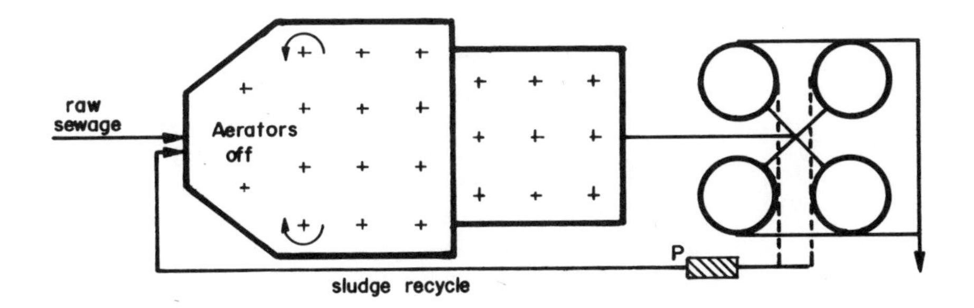

FIGURE 11. PLAN OF ALEXANDRA PLANT, JOHANNESBURG

shows results before and after the modification.

TABLE 1. COMPARISON OF EFFLUENT QUALITY BEFORE AND AFTER DENITRIFICATION

	BOD	COD	TKN	N/NO_3	Loss of N	P	Alka-linity	pH
Feed	420	600	37,6	-		6,3	290	7,1
Effluent	10	83	5,2	22	27,7%	6,3	44	7,1
Feed	340	610	37,4	-		6,7	190	7,1
Effluent	7	50	4,4	0,9	85,8%	6,7	150	7,6

ALL RESULTS IN mg/ℓ WHERE APPLICABLE. DESIGN CAPACITY 27 Mℓ/d.

Experimentation at this plant was stopped due to the overloaded conditions. After the modification to the plant the effluent became extremely clear and the settling rates of the sludge in the final clarifier improved markedly. Figure 12 shows settling rates obtained from this plant compared with those claimed for pure oxygen plants. Phosphorus experiments at this plants indicated that the creation of a fermentation zone would result in high removals of phosphorus.

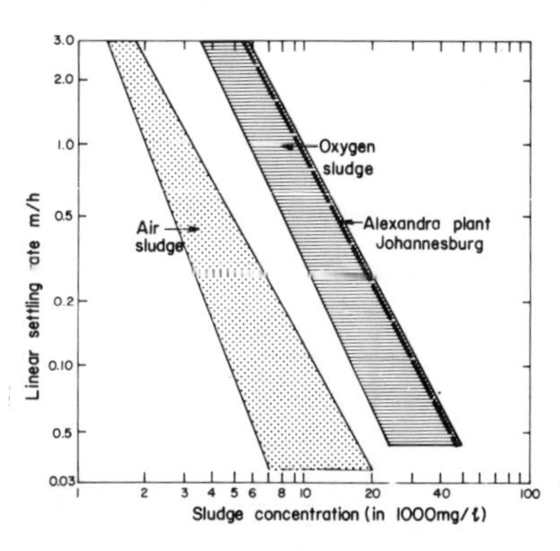

FIGURE 12. LINEAR SETTLING RATES OF SLUDGE

Olifantsvlei plant, Johannesburg

The activated sludge extension of previously existing trickling filter plants consisted of raw waste extended aeration plants comprising four modules, each being able to treat

20 Mℓ/d (5.3 MGD) giving a total of 80 Mℓ/d. At present the four modules receive a constant flow of 140 Mℓ/d (37 MGD) with peak flows being diverted to the trickling filter plant. The arrangement of the aerators is shown in Figure 13.

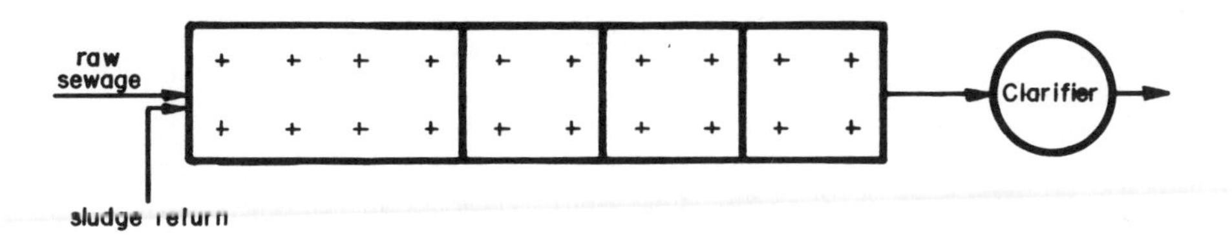

FIGURE 13. LAY-OUT OF LIFANTSVLEI PLANT, JOHANNESBURG

Experimentation on several modules were started by switching off surface aerator units near the influent to the aeration basin. This resulted in the removal of about 75% of the total nitrogen in the system. The interest in phosphorus removal in switching off of more surface aerator units until fermentation conditions developed at the head end of the plant. This caused incomplete nitrification but since ammonia did not effect the phosphorus removal mechanisms, the ammonia concentration was allowed to reach 5 mg/ℓ while keeping the nitrate level below 2 mg/ℓ. This was found to be essential for the removal of phosphorus. The plant had no DO or other controls and it was extremely difficult to balance the oxygen input with the load. Nevertheless it was possible to operate this plant to produce average total phosphorus concentrations of less than 1 mg/ℓ.

Daspoort Pretoria

This plant shown in Figure 14, is again an extension of

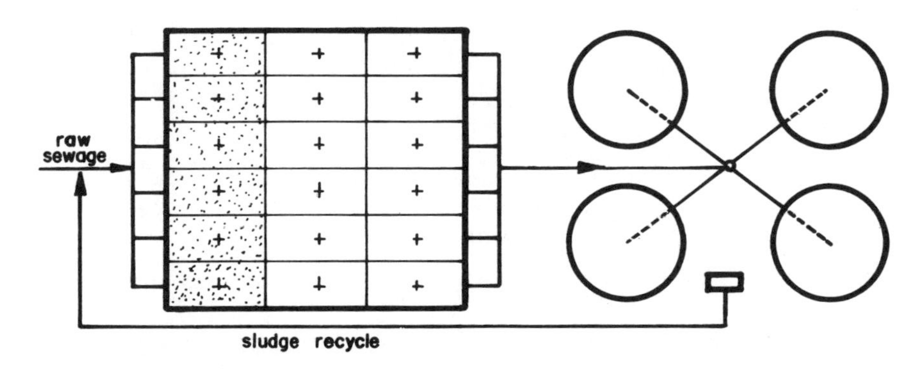

FIGURE 14. LAY-OUT OF DASPOORT PLANT, PRETORIA

an existing trickling filter plant. The plant was designed to handle 30 Mℓ/d (8.4 MGD). Unlike Olifantsvlei this plant had only 3 aerators per run with 6 runs in parallel. Switching off of the first aerator during the present period when the plant is not fully loaded resulted in extremely good phosphorus removal averaging 0,5 mg/ℓ at the cost of some ammonia conversion. As in Olifantsvlei the method is to keep the nitrates low at the cost of allowing some ammonia in the plant effluent. Unlike Olifantsvlei this plant treats a highly variable feed of settled sewage. Influent nitrogen varies between 0.2 and 2.8 times the average value. The plant has no controls other than hand operated overflow weirs.

Brasilia

One of the two plants of the City of Brasilia was modified to allow for 2 anoxic basins (shaded in Figure 15) with temporary recycling from the aeration to the first anoxic basin. The plant suppliers did not take cognisance of the fact

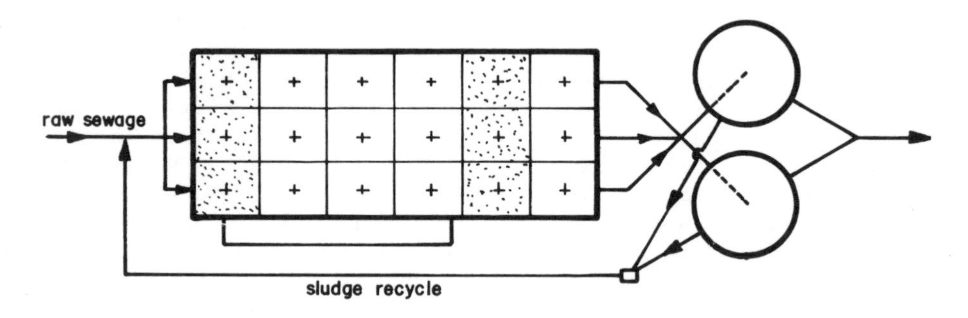

FIGURE 15. LAY-OUT OF BRASILIA NORTH PLANT

that the City was 1 000 m above sea level with the result that the actual oxygenation efficiency is well below the designed value. Good nitrogen removal was hampered by the lack of oxygen, but it was possible to reduce the phosphorus to less than 0,5 mg/ℓ. Removal was not continuous due to the effect of too low dissolved oxygen on the mechanism for uptake of phosphorus.

Experiments were conducted at a number of other smaller plants in South Africa and where-as some very good results were obtained the lack of control lead to sporadic results.

PLANTS DESIGNED ON THE BARDENPHO MODEL

About 20 plants based on this flow scheme are either under construction or have been completed in South Africa. The largest of these is the Goudkoppies plant in Johannesburg presently being commissioned, the Cape Flats Plant in Cape Town presently under construction and the Johannesburg Northern Plant also presently under construction. All three of these plants were designed to treat 150 Mℓ/d (40 MGD) each. All three are similar in that they have a fermentation zone for the conditioning of the sludge for phosphorus removal, plus two anoxic zones for the removal of nitrogen. All three plants have primary sedimentation with separate sludge treatment for the primary and secondary sludges. A lay-out of the Johannesburg Goudkoppies plant is shown in Figure 16. These three plants will all be fully loaded when they come on line and the first results from the Goudkoppies plant should be

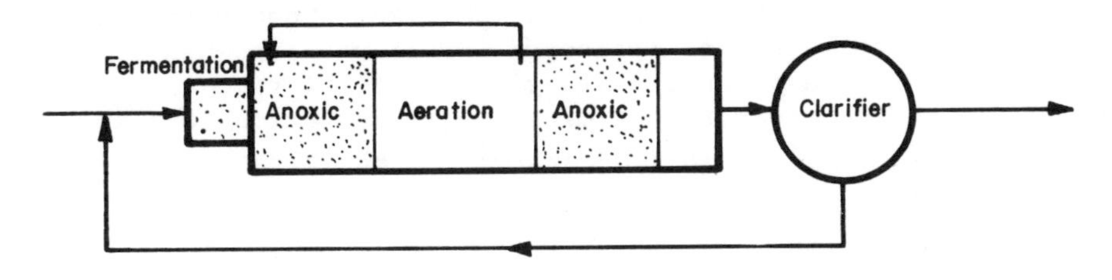

FIGURE 16. SCHEMATIC PLAN OF GOUDKOPPIES PLANT, JOHANNESBURG

available within two months. The following is a discussion of smaller Bardenpho plants under operating conditions.

Secunda, Transvaal

This plant was constructed to serve a mushroomed dormitory town that develop around the large coal to gasoline conversion plant under construction. Tenders called for a design and construct bid and the successful tenderer based his aeration basin on the concept of Alexandra as can be seen in Figure 17.

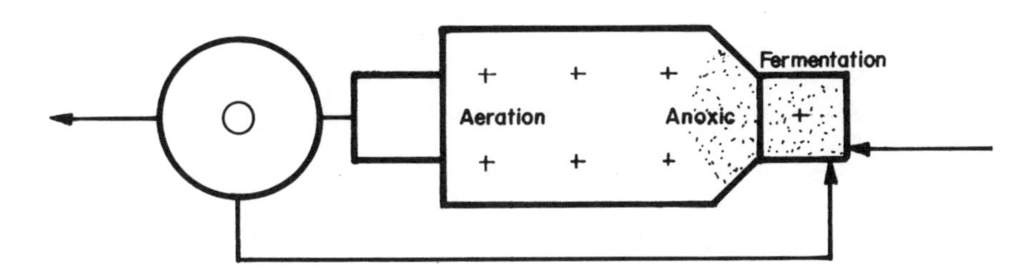

FIGURE 17. LAY-OUT OF SECUNDA PLANT

A fermentation basin was added to the influent side of the plant while a small aeration basin was added between the main aeration and the clarifier in order to ensure sufficient dissolved oxygen in the clarifier so as to minimize re-solubilization of the phosphorus in the sludge. This plant could be considered a hybrid plant and the results tended to confirm this. When operated for maximum nitrogen removal the total effluent nitrogen level could be kept below 2.5 mg/ℓ. The effluent was further in a maturation pond and the average nitrogen concentrating leaving the pond was less than 2 mg/ℓ. The single stirrer in the fermentation basin tends to create a vortex and introduce oxygen to the fermentation basin. No phosphorus release took place in the fermentation basin and this was reflected in the absence of phosphorus removal in the plant. In order to improve conditions the operator cut back on his total oxygen input and the ammonia concentration of the plant effluent has increased. This was accompanied by a reduction in the phosphorus to give about 70% removal. Plans are now underway to optimize the fermentation zone in order to reduce the phosphorus to below 1 mg/ℓ.

Meyerton, Transvaal

This plant incorporated all aspects of the Bardenpho process, and is similar to the Goudkoppies plant. The plant was designed to treat 6 Mℓ/d (1.5 MGD). Only half the town has been sewered presently, the remaining half being served by suction tanks. The sudden discharge of these tanks leads to unstable conditions at the plant. Between 80 and 90% nitrogen removal is obtained and between 60 and 70% phosphorus removal. This plant will presently be studied by my firm in order to optimize the removal of nutrients.

Mitchells Plain, Cape Town

This plant was designed to treat 8 Mℓ/d. The Plant was designed with two parallel runs of which one run has the full compliment of basins according to the Bardenpho process. Some fundamental errors were made in the design of this plant such as the installation of stirrers imparting more than 40 W/m^3 to the basin content compared with the less than 4 W/m^3 required for this function. Aeration is by means of fine bubbled diffused air, the air being supplied by Roots blowers. The plant is only 10% loaded and difficulty is experience in tailoring the oxygen supply to the requirements. The results are sporadic but total phosphorus values in the effluent of less than 0.5 mg/ℓ are being measured regularly.

The variation in the phosphorus removal efficiency could be attributed to the lack of organic matter entering the plant for the complete destruction of nitrates formed. The effluent from this plant however, is very clear with suspended solids concentrations of less than 10 mg/ℓ and BOD values well below 10 mg/ℓ.

EFFLUENT QUALITY

During the first stage of the development of the process, emphasis was placed on the maximum removal of nitrogen. Control of the oxygen and the recycle rate showed that it was possible to control the effluent nitrogen to about 3 mg/ℓ, made up of less than 0.3 mg/ℓ of ammonia nitrogen, about 0.7 mg/ℓ of TKN and 2 mg/ℓ of nitrate and nitrite. This represented some 92% removal of nitrogen. The later development of phosphorus removal and the finding that most of the nitrogen will disappear in the streams has lead to the neglect of high nitrate removals except insofar as it influenced the removal of phosphates. It was assumed by some researchers that high removals of both would not be compatible, but his notion was disproved.

From all the experience to date, it is essential that the nitrate level be kept low to ensure phosphate removal. In the modified plants this is presently done by cutting back on the aeration, even though this means that incomplete nitrification will take place. The ammonia leaving the plant, never more than 3 to 4 mg/ℓ can be removed by post chlorination, but new plants designed on the Bardenpho model will make provision for toning down the aeration without enhancing the release of ammonia in the effluent. Ammonia nitrogen concentrations will be less than 0.5 mg/ℓ and this could conveniently be removed by post chlorination.

It was also found that during warm weather, sandfilters will remove 2 to 3 mg/ℓ of nitrate nitrogen plus 0.2 to 0.3 mg/ℓ TKN. With the addition of some glucose or methanol, the nitrate removal could be increased. Therefore while it is not expected that the total effluent nitrogen level of the Bardenpho plant operating under conditions such as in Florida will be much below 4 mg/ℓ, this level could be reduced to below 2 mg/ℓ by chlorination and sand filtration with or without the addition of a small amount of chemicals.

The biological removal of phosphorus is a positive response to the creation of fermentation conditions somewhere in the biological reactor. Experience on numerous installations, both full-scale and pilot plants have demonstrated the direct response of the activated sludge system to the inducement to remove phosphorus through the provision of fermentation conditions. It was possible to reduce the phosphate content of the effluent to below 1 mg/ℓ before sandfiltration in a variety of plants by creating the necessary pre-conditions. In a properly designed plant, it is expected that the effluent concentration could be reduced to less than 0.5 mg/ℓ on a regular basis with some further removal taking place on sandfilters.

The longer retention times also result in more complete removal of the organic matter, resulting in both a low remaining BOD_5 value of less than 5 mg/ℓ which partly results from improved removal of suspended solids, and in better removals of non-degradable organic matter as evidenced in improved removals of the COD. The expected effluent quality of the process, with chlorination but without sandfiltration will thus be as follows:

BOD_5	less than 5 mg/ℓ
COD	15 to 40 mg/ℓ depending on initial strength
Organic N	0.7 to 1 mg/ℓ depending on the initial strength
Nitrates and nitrites	2 to 3 mg/ℓ
Ammonia nitrogen	trace
Suspended solids	less than 10 mg/ℓ
pH value	above 7 for neutral soft water
Phosphorus as P	0.5 to 1 mg/ℓ

FAILSAFE BACKUP SYSTEMS

All systems for the removal of nutrients by the addition of chemicals rely on a back-up system for reliability. The more reliability expected, the more elaborate the back-up system. Provided toxic substances could be excluded from the sewage, the biological system is more reliable in terms of overall removal than would be a system for chemical addition without a backup system.

For a higher degree of reliability of nutrient removal in the Bardenpho system, back-up systems could be provided for the precipitation of phosphorus with Alum or Iron salts, or for the addition of a carbon source to the second anoxic basin for the removal of the remaining nitrates.

COST CONSIDERATIONS

Even though the process is not new in the sense that it uses a combination of reactions observed signly or in some combinations in many existing plants and even though the process has been operated in many forms over the last number of years in many countries, the first question that seems to arise when discussing the process is the risk involved. When discussing the risk one must examine the other alternatives and the costs of these alternatives.

The only alternative worth considering for the removal of nigrogen is the two or three sludge system for carbon removal and nitrification, followed by denitrification with methanol as depicted in Figure 18. With the same sewage as feed and

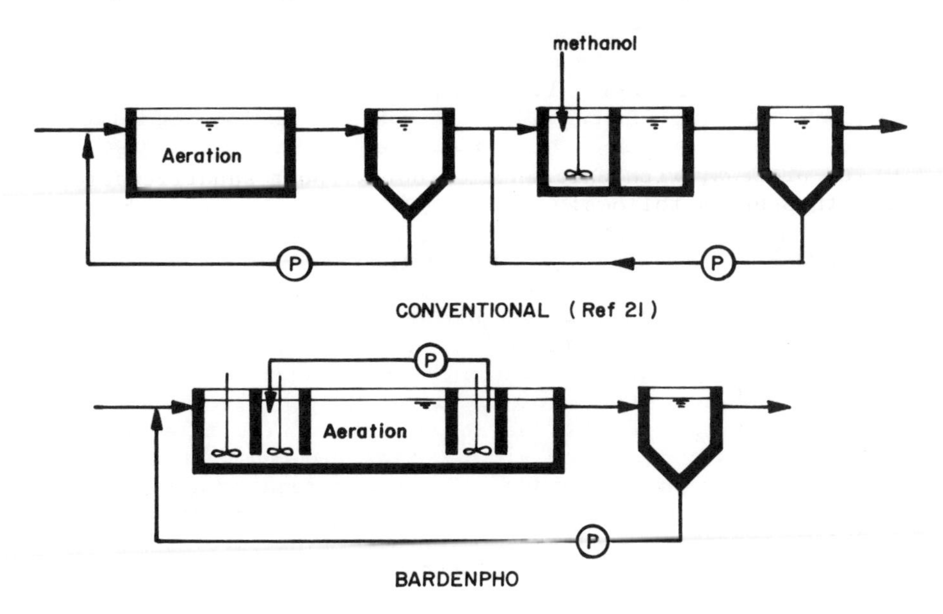

FIGURE 18. COMPARISON OF CONVENTIONAL WITH BARDENPHO PROCESS

the same solids retention time, nitrification in one plant would be accompanied by nitrification in the alternative plant and no risk would be involved. It is inconceivable that no denitrification will take place in the Bardenpho plant since the same mechanism is responsible for denitrification in the alternative plant. The only risk now is that the removal on nitrates in the Bardenpho system will not be as high as that in the alternative since a different substrate is used as hydrogen donor. However, should a problem arise - which is doubtful - this could be rectified by the addition of some chemicals to the second anoxic basin, the quantity being consederably less than for the alternative scheme. In the alternative scheme there is furthermore the risk that low alkalinity may result in the lowering of the pH value with nitrification without internal denitrification.

The only alternative to the biological removal of phosphorus worth considering is the addition of chemicals such as Alum to the mixed liquor. The Bardenpho process does not exclude this option and the only risk to the plant owner consist of the provision of some capacity to provide for the fermentation zone. Later cost comparisons were made for a 1 MGD plant for which the cost of the fermentation zone would be of the order of $20 000 while the annual saving in chemicals would be of the same order.

When comparing the plant and operating costs of the Bardenpho process with the alternatives, only elements that differed were considered. It is interesting to note that the cost of the aeration basin itself for a recently completed 40 MGD plant was 17% of the total cost of the plant, the rest of the cost being accounted for by the ground work, terracing, primary and secondary tanks, pipework, head of works, electricity supply, switchgear, aerators, pumps and for roads and drainage.

The assumptions for the design of the two systems are shown in Table 2. The retention times required for the various basins are shown in Table 3. It should be noted that the cost per unit volume of clarifiers can be from 2 to 3 times higher than that of aeration basins. The mixed liquor recycle pumps consists of simple axial flow pumps for large volumes at low pumping heads. The power requirements for the alternatives are shown in Table 4. The reduced power requirement for aeration results from the use of nitrates as electron acceptor in the place of oxygen.

Some provision was made for the addition of chemicals to the Bardenpho for back-up during emergencies.

TABLE 2. COMPARISON ASSUMPTIONS

Anoxic mixing Energy	5 W/m^3 (0.2 $hp/1000\ ft^3$)
Methanol	11c/kg (5c/1b)
Alum dosage Al : P ratio	1.8:1
Power cost	4c/kWh
Aeration efficiency (std. cond.)	1.8 kg O_2/kWh (3 lb/hp.hr)
Operating cost period	20 years
Interest rate	7%

TABLE 3. DESIGN BASIS

BOD_5	=	200 mg/ℓ
SS	=	200 mg/ℓ
TKN	=	30 mg/ℓ
Phosphorus	=	11 mg/ℓ
Minimum temperature	=	15°C
Maximum MLSS	=	4 000 mg/ℓ
Nitrification SRT	=	11.5 days

TABLE 4. COMPARISON OF BARDENPHO AND CONVENTIONAL SYSTEMS DETENTION TIME

Basins	Detention time (hrs)	
	Bardenpho	Conventional
Fermentation	1.5	-
Anoxic first stage	2.5	-
BOD removal-nitrification	11.5	11.5
Clarifiers		3.1
Denitrification	2.5	1.5
Post aeration	1.0	1.0
Clarifier	3.1	3.1
	22.0	20.2

Relative clarifier unit cost 2 to 3 times unit cost of aeration basin.

TABLE 5. ADDITIONAL EQUIPMENT

Bardenpho	Conventional
Stirrers for 3 basins	Stirrers for one basin
Mixed liquor recycle pumps	Sludge recycle pumps
	Clarifier equipment
	Pump Station
	Pipework
	Controls

TABLE 6. POWER REQUIREMENTS

Basins	Bardenpho	Conventional
	kW/MGD	kW/MGD
Fermentation	1.5	-
Anoxic - stage I	2.5	-
BOD removal - nitrification	42	53
Mixed liquor recycle	1.0	-
Clarifier	-	0.5
Sludge recycle	-	4.0
Denitrification	2.5	2.0
Post aeration	4.0	4.0
Secondary clarifier	0.5	0.5
Sludge recycle	4.0	4.0
	58 kW	68 kW

TABLE 7. PRESENT WORTH OF POWER AND CHEMICAL COST FOR BARDENPHO VS CONVENTIONAL

	Bardenpho ($/MGD)	Conventional ($/MGD)	Savings ($/MGD)
Power	133 600	156 600	
Methanol	10 000	159 400	
Alum for P removal	15 000	322 700	
Total cost without P removal	143 600	316 000	172 400
Total	158 600	628 700	480 100

ADVANTAGES OF BARDENPHO PROCESS

1. The Bardenpho process has other process advantages over the conventional two or three stage process in that less alkalinity is destroyed, resulting in better overall performance. In soft water areas, it would be impossible to operate the conventional system without the addition of alkalinity to the mixed liquor since the drop in the pH value will result in incomplete nitrification and the deflocculation of the sludge and in increase in the suspended solids in the effluent. Addition of Alum would further reduce the alkalinity. In such instances, it would be advisable to combine the nitrification and carbon removal steps of the conventional process and add a first stage anoxic basin for the removal of some of the nitrates formed in order to avoid having to add alkalinity.

2. A second advantage of the system is the finer control allowed by the separation of the zones. By controlling the dissolved oxygen in the mixed liquor of the aeration basin and the recycle rate, the degree of denitrification could be controlled without affecting the nitrification. In channel type systems where a substantial degree of denitrification could be achieved, the control of the system is rather unresponsive to a change in the oxygen input and overcompensation results. Cutting back on the oxygen input also cuts back on the total aerobic zone and on the rate at which the liquid moves in the channels. Accurate control is difficult.

3. The high recycle rates of mixed liquor results in a completely mixed system while the high solids retention time provides a large inventory against minor shock loads.

4. The anoxic zones operate without any input of oxygen. Only organisms that can use nitrate nitrogen as electron

acceptor can compete for food. Full adsorption of all the available COD takes place in the first anoxic basin with subsequent digestion of the material in the aeration zone. The COD of the filtrate of the mixed liquor leaving the first anoxic basin is similar to that of the plant effluent. This means that little feed is available for obligate aerobic bacteria in the aeration basin. Most filamentous organisms are obligate aerobes and cannot survive. This phenomenum was first observed by Krauss when developing his process. Experience with full-scale plants showed that when some denitrification occurred the effluent quality improved as can be seen from the information in Table 1.

5. Simplicity of operation. The main operating parameters are the SRT, dissolved oxygen in the aeration and the reaeration sections and the recycle rates of the mixed liquor. The pilot plant, having no standby equipment, was operated for a period of nine months without supervision except for a visit once a day.

6. Transporting of chemicals is minimized. The removal of nutrients will not be interrupted during interruptions in the supply of the chemicals.

7. The cost of nutrient removal will not escalate with the increase in the cost of chemicals over the lifetime of the plant except with an escalation in the power cost which will be proportionally less than for the conventional system.

8. The phosphates in the sludge can be released and used by plants. The dry solids concentration of the phosphates can reach values of 8% as P which would put the sludge on a par with the best fertilizers.

9. Flotation thickening of the sludge could be carried out without the addition chemicals which would reduce the volume of the sludge for land treatment. Further dewatering is possible with the addition of 2.5 kg/dry ton of solids (5 lb/ton).

ENERGY CONSIDERATIONS

The use of biological systems for the treatment of effluents especially using high SRT's usually result in a concern about the power consumption. Meiring[22] investigated the power requirements for biological treatment plants in South Africa based on cost of electricity of 1,2 c/kWh. On this basis the treatment cost amounted to 26 cents per person per year against about $71 per person per year for normal domestic power and about $50 per person per year for automobiles alone, excluding public transport.

He also investigated the energy component for the treatment of sewage by physical chemical means and concluded that this latter process will consume in total twenty times the amount of energy required for the biological treatment plant.

The process could be applied using raw screened sewage, but when the power cost renders generation of power economical, settled sewage could be used. Since nitrification must be achieved in order to remove the nitrogen, energy can be saved through the use of the nitrates as electron acceptors in place of oxygen.

SUMMARY

The Bardenpho process offers an economical alternative for the removal of plant nutrients from wastewater in a simple, easy to operate biological plant without the addition of costly chemicals or with only minimal chemical addition in case very high removals of the nutrients are required. The process is most competitive in the warmer climates but will perform well when the mixed liquor temperature remains above 10°C. With some chemical addition, effluent standards can be obtained similar to those obtainable by the conventional process but with a considerable saving in chemicals.

REFERENCES

1. Barnard, J.L. "Biological Denitrification" Jour. Inst. Wat. Poll. Contr. Fed. Vol. 72 No. 6, 1973.

2. Ludzack, F.J. and Ettinger, M.B. "Controlled operation to minimize activated sludge effluent nitrogen." J. Wat. Pollut. Control Fed., 34, (9), September 1962, 920.

3. Balakrishnan, S. and Eckenfelder, W.W. Jr. "Nitrogen removal by modified activated-sludge process." J. sanit. Engng Div. Am. Soc. civ. Engrs., 96, 1970, SA2.

4. McCarty, P.L., Beck, L. and St. Amant, P. "Biological denitrification of wastewater by addition of organic materials." Proc. 24th ind. Waste Conf. Purdue Univ., 1969, 1271.

5. Johnson, W.K. and Schroepfer, G.J. "Nitrogen Removal by Nitrification and Denitrification", Journal of the Water Pollution Control Federation, Vol. 36, No. 8, 1964, p. 105.

6. Stern, L.B. and Marais, G.v.R., "Sewage as an Electron Donor in Biological Denitrification:, Research Report No. W. 7, Dept. of Civil Eng., University of Cape Town, Republic of South Africa (1974).

7. Barnard, J.L. "Cut P and N Without Chemicals", Water and Wastes Engineering, 11, 7, 33 (1974)

8. Sutton, P.M., Murphy, K.C. and Jank, B.E. "Kinetic studies of single sludge nitrogen removal systems". Progress in Water Technology. Vol. 10 No. 1/2.

9. Wuhrmann, K. Nitrogen removal in sewage treatment processes. Verh. int. Ver. Limnol., 15, 1964, 580.

10. Venter, S.L.V., Halliday, J. and Pitman, A.R. "Optimisation of the Johannesburg Olifantsvlei extended aeration plant for phosphorus removal". Progress in Water Technology. Vol. 10. No. 1/2.

11. Vacker, D., Connell, C.H. and Wells, W.N. "Phosphate removal through municipal wastewater treatment at San Antonio, Texas", Jour. Wat. Poll. Contr. Fed. Vol. 39 No. 5, May, 1967.

12. Milbury, W.F., McCauly, D. and Hawthorne, C.H. "Operation of conventional activated sludge for maximum phosphorus removal", Jour. Wat. Poll. Fed. Vol. 43 No. 9, September, 1971.

13. Garber, W.F. (1972). "Phosphorus removal by chemical and biological mechanisms." Applications of New Concepts of Physical Chemical Waste Water Treatment, Vanderbilt University Conference Sept. 1972, Pergamon Press.

14. Levin, G.V., Topol, G.J., Tarnay, A.G. and Samworth, R.B. (1972). Pilot plant tests on phosphate removal process. Journal Water Poll. Control Federation. 44. (10).

15. Harold, F.M. "Enzymic and genetic control of poly-phosphate accumulation in A. aerogenes" J. Gen. Microbiology 35, 81 - 90 (1964).

16. OSBORN, D.W. and Nicholls, H.A. "Optimization of the activated sludge process for the biological removal of phosphorus". Progress in Water Technology. Vol. 10 No. 1/2.

17. Downing, A.L., Painter, H.A. and Knowles, G. "Nitrification in the activated-sludge process." F. Proc. Inst. Sew. Purif., 1964, (2), 130.

18. Marais, G.v.R. Personal Communication.

19. Pitman, A.R. "Bioflocculation as a means of improving the dewaterability of activated sludges". Wat. Poll. Control. Vol. 74, 1975, No. 6.

20. Hemstead, T. and Marais, G.v.R. "Theory of Aerobic Digestion". Unpublished paper.

21. Barth, E.F., Brenner, R.C. and Lewis, R.F."Chemical-biological control of nitrogen and phosphorus in wastewater effluent." J. Wat. Pollut. Control Fed., 40, (12), December 1968, 2040.

22. Meiring, P.G.J. "Energie behoeftes by die biologiese behandeling van rioolwater" Chemsa, July, 1977.

CHAPTER 6 PART II

USE OF THE BARDENPHO PROCESS WITH LIME PRE-TREATMENT

The treatment of raw sewage with lime for the production of a high quality effluent was suggested by Horstkotte et al.[1] This followed tests of the so-called independent physical chemical treatment scheme. Horstkotte et al showed that lime treatment of raw sewage ahead of biological treatment had a number of advantages some of them being

1. the removal of suspended and colloidal organic material requires a much shorter biological retention time as well as a lower power consumption,

2. the recarbonation of the lime-treated liquid is effected by CO_2 developed in the biological system. The addition of CO_2 is a costly process in the independent physical process, but it can be obtained as a by-product in the biological process. The effluent from the biological system will then have good stability,

3. nitrification will not destroy the alkalinity and since optimum pH values for nitrification is 8.2 the pH will stabilize around this value giving optimal conversion of ammonia to nitrates,

4. the clarity of the effluent is very good and a high quality effluent is produced, suitable for most industrial applications,

5. the lime treated raw waste can be segregated in centifuges, the lime re-calcined and re-used after further classification.

In the process as outline by Horstkotte et al and as applied at Contra Cost in California, and at Canberra in Australia, ammonia was converted to nitrate in the first activated sludge system, followed by a second activated sludge unit for denitrification with methanol, as shown in Figure 1. My present knowledge is that both these plants will have the capability of removing nitrates, but due to the high cost of methanol, the nitrates will not be removed.

During lime treatment of raw sewage about 70 to 75% of the organic carbon is removed if the sewage is relatively fresh while less than 20% of the ammonia of TKN is removed. This drastically changes the ratio of organic carbon to nitrogen in the remaining effluent. About 3.5 mg/ℓ of BOD will be required for the removal of 1 mg/ℓ of nitrogen in the first

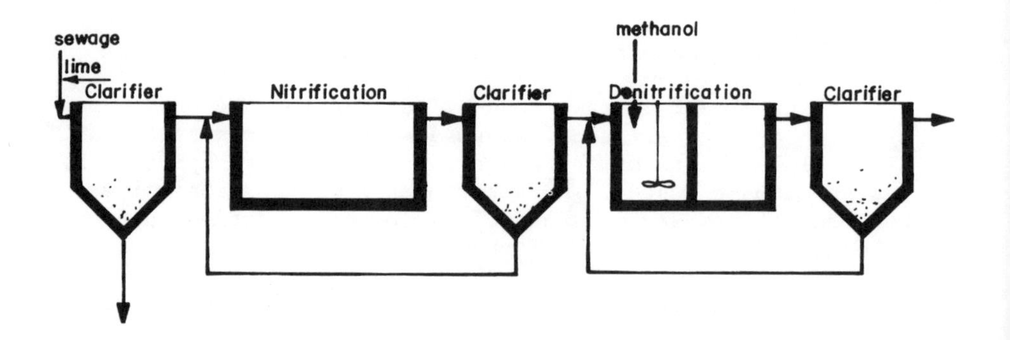

FIGURE 1. PROCESS DESCRIBED BY HORSTKOTTE ET AL

anoxic zone of the Bardenpho process. However, since some of the nitrogen is removed by endogenous respiration, the actual requirement for BOD will be lower. It would therefore be possible to apply the Bardenpho process to lime pre-treated sewage to make optimal use of the remaining organic carbon for denitrification, thereby reducing substantially the requirement for additional methanol for denitrification.

The problem of having insufficient organic carbon for denitrification after lime treatment is peculiar to countries having a high water consumption and specifically to plants where the collector system is not extensive. In countries having a lower per capita water consumption resulting in a stronger municipal waste, or where other conditions such as extensive collector systems and pumping mains result in some fermentation of the sewage before reaching the treatment plant, volatile acids are formed that will not be removed by lime. Rebhun et al[2] reported that the lime treatment of strong municipal waste water resulted in virtually no reduction in the dissolved volatile acids in the raw sewage and that in fact virtually no reduction took place even after sandfiltration and activated carbon treatment, as can be seen from the results of their work in Table 1. From the table it can be seen that the remaining ammonia having a concentration of 60 mg/ℓ could easily be removed by the remaining BOD_5 of 181 mg/ℓ plus allowance for endogenous respiration.

Tests carried out with lime pre-treated settled sewage at Daspoort in Pretoria, indicated that virtual complete removal of nitrogen was possible without the addition of any external carbon source. The convertion of ammonia to nitrates was so complete that the auto analizer would not pick up the results. Nitrate concentrations averaged 1 mg/ℓ. These tests were carried out using a feed that was collected at a certain time of

TABLE I. DIRECT PHYSICO-CHEMICAL TREATMENT OF STRONG MUNICIPAL WASTEWATER. REMOVAL OF ORGANICS. (REF. 2)

CONSTITUENTS mg/ℓ	RAW		CHEMIC. CLARIF.	FIL-TRATION	ACTIVE CARBON
	TOTAL	FILTERED			
SUSP. SOLIDS	450		5	2	2
BOD_5	423	206	181		135
COD	1170	476	366	343	223
PROTEINS	152	63.2	45.5	43	4.5
CARBOHYDRATES	95.5	13.5	4.5	4.5	0.6
VOLATILE ACIDS		210	208	208	193
GREASE & FATS	162	-	nil	nil	nil
DETERGENTS		17	8.4	8.4	0.05
PHOSPHATES PO_4	46	41	0.2	0.2	0.2
COLOR C.U.		194	43.5		3
AMMONIA -N		65	60	60	60

day that would give a representative organic carbon to nitrogen ratio. In this instance the collector system is relatively small and the average retention time in the system was calculated to be around 2 hours. It is expected that with systems having a longer overhaul retention time and where pumping mains and inverted syphons will produce a certain amount of volatile acids, the process would perform even better.

The provision of sufficient volatile acids to ensure the denitrification of nitrates formed in the Bardenpho process can be obtained by the simple mechanism of allowing some acid fermentation to take place in cases where the raw sewage is fresh or the sewage is weak. This could easily be achieved through the flow diagram shown in Figure 2. By allowing for large and deep primary sedimentation tanks, in which primary sludge can be thickened, and allowed to age, sufficient volatile acids can be produced that will not be removed by subsequent lime treatment and will therefore be available for denitrification in the biological system without imparing the quality of the effluent in any way or without effecting the advantages of the lime pre-treatment system. The waste raw

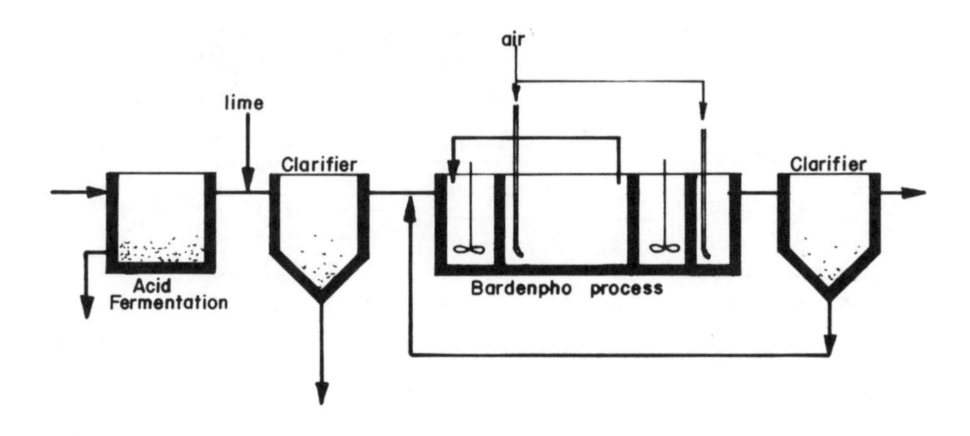

FIGURE 2. ACID FERMENTATION FOLLOWED BY LIME TREATMENT AND BARDENPHO PROCESS

sludge withdrawn from the primary tank would then be in a thickened condition ready for further digestion which would also have the advantage of gas production for the conservation of energy. The flow scheme as outlined in Figure 2, will have the further advantage that a smaller amount of heavy organic material will be mixed with the lime sludge and during the classification process a smaller amount of organic residue will be classified with the lime sludge. This would improve the quality of recalcined sludge.

Another variation of this flow scheme would be to pass some of the raw sewage through an anaerobic zone where acid fermentation will take place. Followed by mixing of this effluent with the raw sewage before lime treatment.

In summary, in addition to the advantages claimed by Horstkotte et al, the following advantages will accrue to the proposed flow diagram:

1. The use of methanol for denitrification will be eliminated or reduced to a minimum.

2. Digestion of primary sludge will result in gas production to offset the total energy cost.

3. The lime sludge will contain less heavy organic material resulting in a better classification in the centrifuge and a better quality re-calcined lime.

REFERENCES

1. Horstkotte, B.A. Niles, D.G., Parker, D.S. and Caldwell, D.H. "Full-scale testing of a water reclamation system." Proc. 45th ann. Conf. Wat. Pollut. Control Fed., Atlanta, Ga, 1972.

2. Rebhun, M. and Narkis, N. "Physico-chemical treatment of raw wastewater : distribution of organics and treatment efficiency". Presented at 7th IAWPR Conference, Paris 1974.

7. PHOSTRIP®*: A BIOLOGICAL-CHEMICAL SYSTEM FOR REMOVING PHOSPHORUS

L. C. Matsch, Union Carbide Corporation, Linde Division, Tonawanda, New York

R. F. Drnevich, Union Carbide Corporation, Linde Division, Tonawanda, New York

ABSTRACT

The PHOSTRIP system for removing phosphorus from waste waters is a combination process. First the biological phenomenon of phosphorus uptake and release by bacteria under oxic and anoxic conditions respectively is used to concentrate the phosphorus to be removed in a small side stream. Subsequently the phosphorus is precipitated from this side stream by means of lime. The yearly total cost of this process is about half of that of the conventional chemical precipitation process. A combination of the PHOSTRIP system with extensive nitrogen removal by means of denitrification was found to be economically most attractive.

INTRODUCTION

The effect of the release of excessive amounts of nutrients contained in treated wastewaters has been recognized for many years as an important contributing factor to the premature aging of the receiving waters. It has also been shown by many workers that the discharged phosphorus is particularly harmful, and probably more harmful, than nitrogen. For this reason, phosphorus is being removed from wastewater treatment plants in many locations where it was considered to be critical. With the new trend toward upgrading the quality of our rivers and lakes, the nation is faced with the need to remove the phosphorus from many more wastewater streams than before, and it is most important and particularly timely to have an efficient and highly economical process available for this purpose.

*PHOSTRIP is a registered trademark of Union Carbide Corp.

The subject of this paper deals with such a process; namely the PHOSTRIP system. The conventional way of removing phosphorus from wastewaters is by chemical precipitation of phosphates by means of ferric chloride alum and lime, whereby the precipitation, according to the EPA Process Design Manual,[1] can be accomplished in the primary clarifier or as part of the secondary treatment, or, alternately, as a tertiary treatment process. The use of lime is normally limited to primary clarifier addition or tertiary treatment because precipitation with lime in the secondary system would cause the pH to rise to levels (9 to 12) which are not conducive to the best performance of the activated sludge process.

In contrast to the "straight precipitation", the PHOSTRIP system is a combination of a biological step and a precipitation step. The purpose of the biological step is to remove the phosphorus from the wastewater by inducing the activated sludge bacteria to take it up and release it into a small sidestream at several times higher concentrations in the form of orthophosphate. The second step consists of precipitating the phosphate from this sidestream by means of lime. The reason for using lime in the removal step is that precipitation by lime is pH dependent and not a stoichiometric process, and, therefore, the amount of chemical needed is approximately proportional to the liquid flow rate and almost independent of the phosphorus concentration. In the case of the PHOSTRIP system, the biological step concentrates the phosphorus in a stream of the size of about 10-15% of the influent flow rate, and therefore the amount of lime needed is only about 10-15% of the quantity which would be required to treat the influent stream directly with lime.

The ability of microorganisms to accumulate phosphorus in their bodies is not a new observation. Volutin -- a metachromatic granule composed of polyphosphates -- was first observed by Ernst[2] in 1888. Since that time, microorganisms such as bacteria, fungi, yeasts, algae and protozoa, have been shown to contain intracellular volutin granules.[3-7] Organisms such as _E. coli_ are capable of producing polyphosphate granules under proper environmental conditions.[8] Srinath et al[9] concluded that protozoa of the _Epistylia sp_. could be related to excess phosphorus uptake in activated sludge systems. Feng[10] demonstrated that the operating conditions of the activated sludge process affected the rate and capacity of the process to remove phosphorus.

But to induce the uptake of an "excess" amount of phosphorus by the bacteria in an activated sludge system is, in most cases, not sufficient to reduce the phosphorus content in the effluent to the usually desired low level of <1 mg/l of total phosphorus. Therefore, it is important that Sekikawa et al[11] studied the conditions which were responsible for the release of phosphorus by activated sludge. One of Sekikawa's conclusions was that the lack of dissolved oxygen was a major cause of orthophosphate release. It was Levin[12] who used this information as well as the results of his own studies[13] to design a process which was capable of removing phosphorus from wastewater. He named this system the PHOSTRIP system. It is based on the fact that the microorganisms under aerobic conditions -- in the aeration basin -- are taking up more phosphorus than they normally need, while under anoxic conditions the excess phosphorus is released back into the liquid phase, in the so-called "stripper" tank. But before the process itself is described in detail, we will discuss the theoretical background for the phosphorus uptake and release mechanisms.

BIOCHEMICAL MECHANISM

The performance of the PHOSTRIP system is based on the formation and elimination of intracellular polyphosphate granules (volutin). This phenomenon of polyphosphate formation and decay is induced through the continuous cycling of activated sludge through aerobic and anoxic conditions. As a consequence of the aerobic environment, the conditioned microorganisms remove phosphate from solution and produce the storage product, volutin. These organisms then release phosphate into solution while under a condition of stress (anoxic surroundings). The precise biochemical pathways have not yet been delineated, but several hypotheses have been proposed.

Fuhs and Chen[14] isolated organisms capable of excessive phosphorus uptake from PHOSTRIP systems operating at Baltimore, Md. and Seneca Falls, N.Y. Of these isolates, the organisms of the genus *Acinetobacter* most vigorously demonstrated rapid uptake of phosphorus under aerobic conditions and rapid reduction in the size of polyphosphate granules while anoxic. Russ[15] isolated an organism which he named *Acinetobacter phosphadevorus* from the Rilling Road Sewage Treatment Plant, San Antonio, Texas. This organism showed the same phosphorus storage potential as those isolated from the PHOSTRIP systems. It is significant to note that these organisms are obligate aerobes which effectively compete with the facultative species in a system with prolonged anoxic periods. It is hypothesized that the formation of the volutin granules during the aerobic portion of the PHOSTRIP system is largely responsible for this ability to compete.

Investigations into the purpose of polyphosphate formation have indicated that the function of the storage product is either to serve as a phosphate reserve, or to regulate the phosphorus economy of the cell, or, finally, to serve as an energy source.

The first explanation contends that the major purpose of the stored polyphosphates is to supply phosphorus for metabolism during periods of phosphorus starvation. However, this does not seem very likely because the bacteria in the PHOSTRIP system are never starved of phosphorus and, therefore, there would not be any need for storing reserve phosphorus.

The second hypothesis may eventually explain the storage of phosphorus. The mechanistic interpretation of this pathway is just now being developed.

The hypothesis which presently seems to fit most of the observations best is the third one. According to this hypothesis, the use of polyphosphate is as an energy source, either as a "phosphagen"[16-17] or through direct substitution for adenosine triphosphate (ATP)[18] in energy requiring reactions. A broad definition of a phosphagen is a naturally-occurring compound which stores phosphate bond energy and from which phosphoryl groups can be transferred to adenosine diphosphate (ADP) to produce ATP.

The energy source hypothesis would predict that obligate aerobes in an anoxic environment would utilize energy stored in polyphosphates to continue some of the metabolic functions of the cell. Equation 1 presents an example of a possible use of energy to produce the Krebs cycle precursor, acetyl Coenzyme A, from acetate. The acetate necessary to drive this

$$\text{acetate} + \text{ATP} \rightarrow \text{acetyle Coenzyme A} + \text{ADP} + \text{Pi} \quad (1)$$

reaction is produced by the facultative bacteria while under the anoxic conditions of the stripper tank. Experiments performed at Union Carbide's Tonawanda Labs[19] and Penn State University, where acetates were added to sludges obtained from PHOSTRIP systems, have indicated enhanced phosphorus release while the addition of glucose was significantly less effective at increasing the rate of phosphorus release from the cells. The importance of acetate formation is enhanced by the fact that the organisms of the genus *Acinetobacter* prefer acetates as a substrate and will not metabolize glucose. Since the Krebs cycle cannot function without either molecular oxygen or nitrate as the final electron acceptor, the acetyle Coenzyme A content of the cell increases while the ratio of ATP to ADP decreases. ATP may also be used as the energy source for the formation of other carbonaceous storage products. The low ratio of ATP to ADP will cause the polyphosphokinase enzyme[16] to utilize the bond energy and the phosphoryl group stored in the polyphosphate to produce ATP from ADP. Equation 2, describing this process, is reversible, and the net direction of the reaction is dependent on the relative concentrations of ATP and ADP. This reaction is slow relative to the ATP utilizing reactions; therefore, no ATP buildup is possible during the anoxic period. Alternately, when the ATP level of the cell is diminished as in reaction (1), it is possible

$$ADP + PP_n \rightleftarrows ATP + PP_{n-1} \quad (2)$$

where PP_n = polyphosphate

that short chain polyphosphates replace ATP as shown in equation 3.[7]

$$\text{acetate} + PP_n \rightleftarrows \text{acetyle Coenzyme A} + PP_{n-1} + Pi \quad (3)$$

where P_i = inorganic phosphorus

Either of these reactions may be predominantly responsible for the polyphosphate reduction observed[14,20] in sludges from PHOSTRIP systems held under anoxic conditions. Investigators at Penn State University have also established a significant reduction in ATP levels in sludges leaving the anoxic PHOSTRIP stripper tank.[20]

According to this hypothesis, the sludge leaving the stripper tank may be characterized by the following conditions: (1) low ATP/ADP ratio, (2) high concentrations of acetyl Coenzyme A and/or other carbonaceous storage products, and (3) relatively low polyphosphate levels in the sludge.

The effect of the anoxic period on the polyphosphate level in the sludge is shown in Figure 1. These photomicrographs are sludge samples which were stained to highlight the volutin granules within the bacteria. The results of this test show that the sludge entering the stripper (Figure 1a) contained many large clusters of polyphosphate containing organisms while the same sludge shows a significant reduction in the size and quantity of the polyphosphate granules after only an 8-hour anoxic period (Figure 1b).

Upon return to the aerobic basin of the PHOSTRIP system, the organisms rapidly produce ATP because the reaction rates in the Krebs cycle are enhanced by the low ATP/ADP ratio and the high concentration of the storage products. After a short period of time, these reactions increase the ATP/ADP ratio, thus driving the reaction in equation 2 from right to left. According to this mechanism, the uptake rate of phosphorus in the aeration basin should be zero order with respect to phosphorus concentration in solution. Experiments performed at Union Carbide indicate zero order uptake rate at phosphorus levels greater than 0.5 mg orthophosphorus/liter of solution.[21]

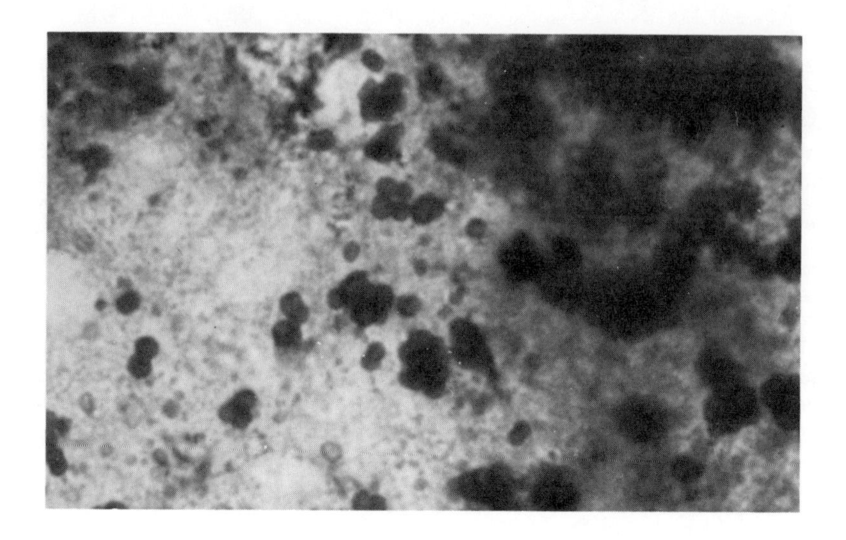

FIGURE 1a.

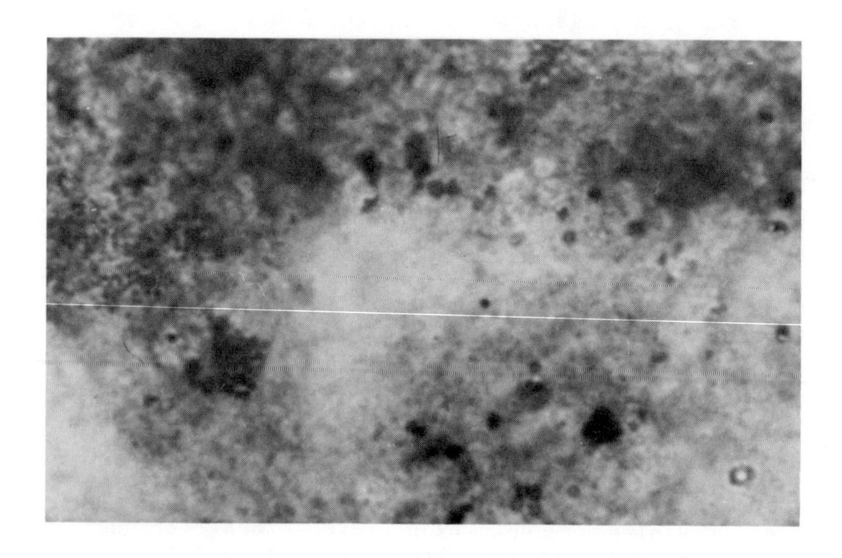

FIGURE 1b.

VOLUTIN GRANULES IN BACTERIA (100X)
ENTERING (a) AND LEAVING (b) THE ANOXIC
STRIPPER TANK

Consistent with the above hypothesis, early testing performed by Biospherics, Inc.[22] and later studies[23] have indicated that phosphate release is inhibited in the stripper tank by low levels of nitrate and dissolved oxygen. The presence of nitrate and oxygen stimulates the ATP producing reactions of the Krebs cycle.

PROCESS DESIGN

As was mentioned earlier, the PHOSTRIP system as designed by Levin[12] is based on the idea of using biological means to concentrate the phosphorus to be removed in a small sidestream of the conventional activated sludge process and remove the phosphorus therefrom by precipitation by means of lime addition. The flow diagram depicting this initial scheme is shown in Figure 2.

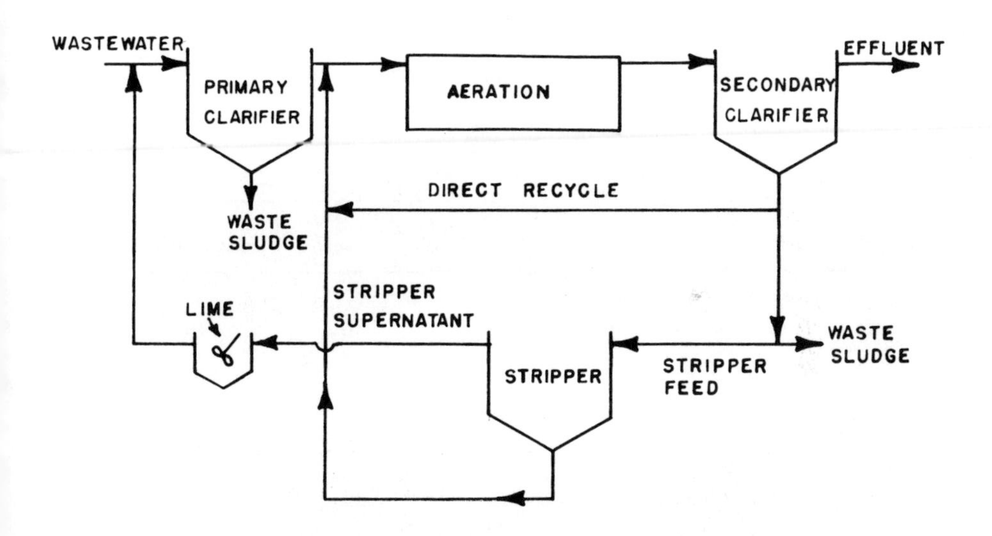

FIGURE 2. GENERAL PHOSTRIP FLOW DIAGRAM

As the diagram shows, the process is designed such that part of the sludge from the clarifier is returned directly to the aeration basin but the other part is fed into a separate tank, the so-called stripper tank. This container is kept anoxic; the sludge stays in it for several hours and gradually makes its way to the bottom from where it is finally withdrawn. This sludge still in an anoxic condition is partially depleted of phosphorus. It is then also returned to the aeration basin where - now exposed to aerobic conditions - it eagerly takes up the phosphorus brought in by the influent wastewater.

The supernatant from the stripper tank, which is enriched in orthophosphate released by the bacteria in the stripper tank, is treated with lime and then fed back into the primary clarifier. If no primary clarifier is foreseen, then the precipitate from the liming vessel can be removed separately and the supernatant fed directly into the aeration basin. It should be noted that in this process design, the stripper tank also has to act as a thickener.

This process was successfully operated on pilot plant scale by Biospherics, Inc. at Rockville, Md. on several different wastewaters. The results of their pilot tests are listed in Table 1.

TABLE 1.

SUMMARY OF PHOSTRIP PILOT PLANT RESULTS

Results of Analysis of Raw Data Obtained from Biospherics, Inc.

	Total Phosphorus		
Location	Raw Waste, mg/l	Effluent, mg/l	Removal, percent
Synthetic Waste (27-day phase duration)	9.6	0.1	99
Washington, D.C. (30-day phase duration)	6.8	0.8	88
Piscatawa, Md. (30-day phase duration)	5.1	0.5	90
Chicago, Ill. (10-day phase duration)	3.0	0.3	90

During the pilot plant testing it was discovered that, since the release of the phosphorus in the stripper tank took several hours, the majority of the released phosphorus remained in the lower part of the stripper and the liquid containing it became trapped in the thickened sludge. The process was therefore modified according to the flow scheme shown in Figure 3; the difference compared to the earlier scheme being that some of the underflow of the stripper was recycled back to the inlet of the stripper. This modification indeed caused a fair amount of dissolved phosphate to end up in the supernatant from where it could be removed by lime addition as before.

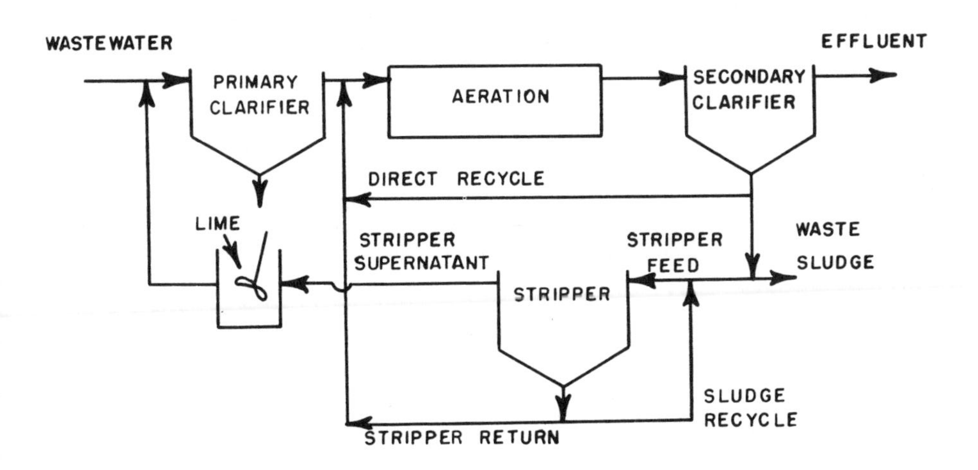

FIGURE 3. SLUDGE RECYCLE OPTION

This improved scheme was used in the full scale demonstration test carried out at the Seneca Falls, N.Y. wastewater treatment plant.[13] This plant had a design capacity of 12,100 m^3/d (3.3 mgd) but was greatly underloaded, handling only about 3800 m^3/d (1 mgd) influent. This condition enabled the plant to handle the full load in one of the two reactor trains, thereby also freeing one primary clarifier for use as a stripper tank during the tests. Although this arrangement was very convenient, the clarifier was much larger than the optimum stripper size, but, of course, could easily handle the recycle sludge. A summary of the results of the tests is listed in Table 2. It can be seen that the plant worked well, removing BOD, SS, as well as phosphorus.

TABLE 2.

RESULTS OF PHOSPHORUS REMOVAL TEST AT SENECA FALLS, N. Y.

Plant Flow, m^3/d (mgd)	3400 (0.9)
Return Flows, percent of raw flow:	
Sludge to Stripper	24
Sludge to Aeration from Stripper	10
Supernatant to Primary Clarifier	14
Total Suspended Solids, mg/l:	
Mixed Liquor	1,440
Sludge to Stripper	7,840
Sludge to Aeration from Stripper	15,910
Influent, mg/l:	
BOD_5	158
Total Phosphorus	6.3
Effluent, mg/l:	
BOD_5	~ 4
Total Phosphorus	.6
Plant Performance:	
BOD_5 Removal, percent	98
Total Phosphorus Removal, percent	90
Lime Dose, Stripper Supernatant, mg/l of supernatant	170
Lime Dose, prorated to mg/l of raw flow	24

Early in 1974 Union Carbide began preliminary testing of the PHOSTRIP system on a pilot scale to determine the feasibility of the process. This pilot plant had a capacity of about 0.95 l/s (15 gpm), using pure oxygen as the source for the dissolved oxygen in the four sequential completely mixed stages. The results were not satisfactory. It was found, for example, that if traces of oxygen or nitrate entered the stripper tank from the clarifier, phosphorus removal suffered. Subsequently, the scheme shown in Figure 3 was applied which improved the conditions but still did not produce consistently high quality effluents. The nature of the process scheme in Figure 3 is such that the supernatant is produced by thickening the sludge in the stripper. Because of solids flux limitations, this mode resulted in the operation of the clarifier without a sludge blanket. Since the sludge recycle scheme makes the stripper approach a completely mixed system, the dissolved oxygen was partially distributed through the tank and somewhat inhibited phosphorus release.

Finally the flow scheme was further modified and operated according to the flow diagram shown in Figure 4. This flow scheme, as discussed by Matsch et al[24] and Drnevich et al[25] utilizes a liquid containing relatively low concentrations of both phosphorus and solids to elutriate the sludge in the stripper tank and, using countercurrent flow, drive the released phosphate into the supernatant. As shown in Figure 4, the preferred elutriant is the supernatant from the lime reactor, but other liquids with similar characteristics; e.g., the final effluent or even the primary effluents (as shown with dotted lines), could also be used.

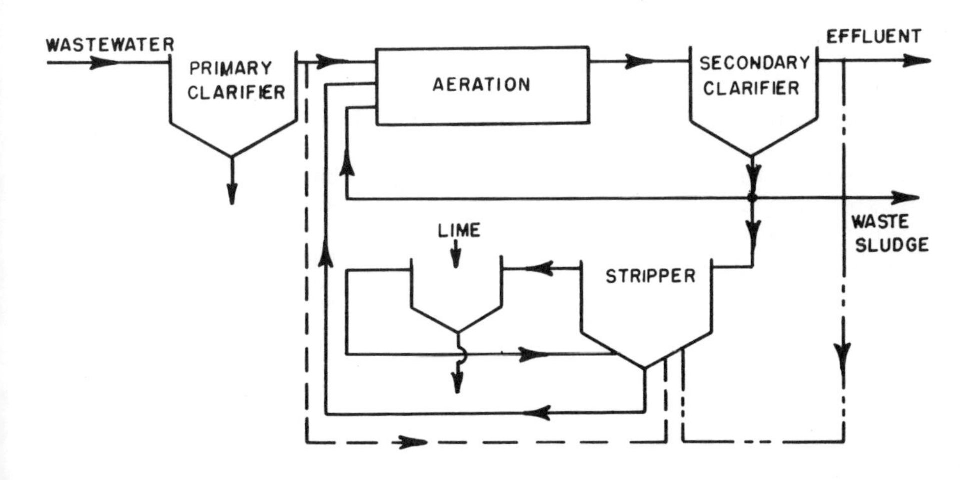

FIGURE 4.
ELUTRIATION CYCLE USING LOW PHOSPHORUS CONTENT LIQUID

This type of operation proved very efficient and reliable, and the results of the Tonawanda pilot plant are listed in Table 3. There are, however, some further significant advantages to this operating mode, but discussion of these requires a short excursion into a related field.

Solids Flux Considerations

A theoretical evaluation of the behavior of settling tanks with respect to thickening has been performed by Dick.[26] Briefly summarizing his considerations, the downward flux rate of the solids, g_t, is the product of the settling velocity and their concentration. Dick found that one can consider the total suspended solids flux as the sum of the flux due to gravity, g_g, and that due to the bulk downward movement of the liquid, g_u, caused by the bottom sludge withdrawal, as expressed in equation 4,

$$g_t = g_g + g_u = c_i v_i + c_i u \qquad (4)$$

where c_i is the initial solids concentration,

v_i is the initial settling velocity,

and u is the liquid velocity due to the underflow.

The relationship between g_g and the solids concentration is somewhat complex because the settling velocity is a strong function of the solids concentration. This latter functionality can usually be expressed well by a logarithmic relationship,[27] and the expression in equation 5 was found very useful.

$$v_i = a \cdot c_i^{-n} \qquad (5)$$

The constants a and n describe the settling characteristics of the specific sludge. This equation applies, however, only in the range of practical interest. Considering the limits of the concentration range, it is clear that at zero solids concentration, the flux is also zero, and it is known that at high concentrations the settling velocity is very low and, therefore, the flux also decreases fast. The general relationship for the entire concentration range is shown graphically in Figure 5.

The effect of the underflow is much simpler, and since u is assumed to be fixed, g_u is proportional to c_i, as also shown in Figure 5.

The total flux g_t is shown graphically as the sum of the two components, and as can be seen, it shows a minimum at an intermediate solids concentration. This minimum value of the flux, called the limiting flux, controls the solids handling capability of any thickening tank.

TABLE 3.

PILOT PLANT OXYGEN ACTIVATED SLUDGE SYSTEM

AT TONAWANDA, N. Y.

Mode of Operation	Precipitated Stripper Supernatant Used for Elutriation	Final Effluent Used for Elutriation
Duration of Phase, days	20	23
Feed Rate, l/s (gpm)	.95 (15)	.95 (15)
Recycle Rate, % of Q	40.	33.
Stripper Feed, % of Q	19.	20.
Stripper Return to Aeration, % of Q	18.	19.
Stripper Supernatant Rate, % of Q	9.	11.
Elutriation Rate, % of Q	8.	9.
Anaerobic Retention Time, hr. (based on stripper return to aeration)	6.1	6.3
Total Suspended Solids, mg/l:		
Mixed Liquor	5,189	4,400
Stripper Feed	18,600	17,500
Stripper Return	18,140	17,540
Effluent	7	10
BOD_5, mg/l:		
Influent	100	120
Effluent	10	18
Total Phosphorus, mg/l:		
Influent	7.5	11.1
Effluent	0.5	0.8
Stripper Return (soluble)	36	60
Stripper Supernatant	58	78
Elutriation Efficiency (c)	.45	.43

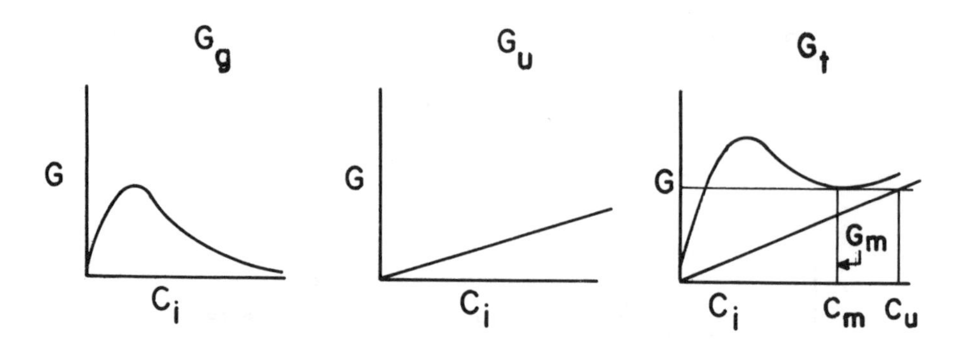

FIGURE 5. SOLIDS SETTLING FLUX OF THICKENERS

Applying the solids flux relationship to the stripper tank it is found that the elutriation mode of operation results in very important advantages. The elutriation system not only reduces the solids feed rate, as compared with the sludge recycle system, but also allows the stripper to be operated without thickening the sludge. These two features result in a very significant reduction in the required size of the stripper tank which in turn is reflected in the overall economics of the process.

Variables Affecting the Process

Some general considerations as to how the phosphorus removal is affected by various design features will round out the description of the process.

The stripper tank is, of course, the focal point of the process, and its design in relationship to the activated sludge process is important. There are two limits for phosphorus removal which need to be observed. The first one is the phosphorus loading capacity of the organisms, and the second, kinetics.

The loading capability of the bacteria is usually not a factor. Not only have phosphorus concentrations of 8 and even 10% of the VSS been observed, but the bacteria normally need to take up and release only a fraction of one percent of their weight as phosphorus per pass in order to achieve the desired performance. These two factors show that the process is inherently capable of easily removing all the phosphorus that can reasonably be expected in municipal applications, with a considerable amount of capacity to spare. As a result only a fraction of the return sludge need be exposed to anoxic conditions in the stripper. The phosphorus balance presented in Figure 6 can be used to illustrate this point. Assuming the feed rate to the stripper to be 15% of the feed rate to the plant - half of the

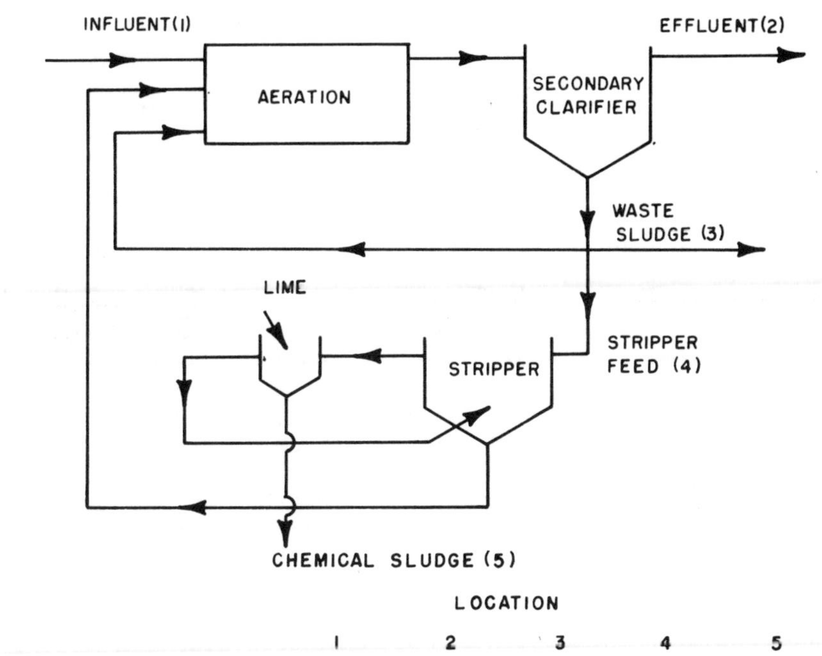

		LOCATION				
		1	2	3	4	5
FLOW RATE	T/D	38000	38000	260	5700	23
TOTAL P	KG/D	380	38	62	1360	280
TOTAL SS	KG/D	3800	114	2600	57000	2300
VOLATILE SS	KG/D	2470	874	2000	43800	1400

FIGURE 6: PHOSPHORUS MASS BALANCE FOR
FOR A 38,000 $m^3$1 (1 mgd) SIZE PLANT

underflow - then at the concentration of 0.77% VSS in the underflow, 43,800 kg VSS/d pass through the stripper. Since 280 kg/d of phosphorus are removed, the change in phosphorus concentration of the VSS needs to be only 0.64%. This is about 21% of the 3.1% phosphorus content of the VSS; that means the bacteria need to give up only about one fifth of their phosphorus during each pass through the stripper.

Concerning the kinetics, we have to consider the uptake rate and the rate of release at phosphorus. The uptake rate of phosphorus has been shown to be zero order with respect to phosphorus concentrations down to very low levels.[22] The general activity of the organisms has, of course, some effect, and, therefore, the F/M ratio will be a factor. In general, the phosphorus uptake rate will increase at increased F/M ratios. Temperature also has a slight effect. Because of the zero order relationship, the uptake rate will be proportional to the retention time in the aeration basin and the number of organisms capable of phosphorus uptake which is related to the MLVSS. The number of these organisms is in turn related to the mass rate of sludge conducted through the stripper.

The performance of the anoxic zone with respect to phosphorus release depends on the kinetics and how close the phosphorus content of the organisms is to their saturation level. Earlier studies have indicated that the rate of phosphate release is dependent on the acetate level of the anoxic solution. This is consistent with Fuhs and Chen[14] who stated that a product was being generated by organisms not responsible for phosphorus uptake that facilitates release in the anoxic zone. The facultative organisms produce acetates and Krebs cycle intermediates during the anoxic period. The rate of formation of these compounds controls the rate of phosphorus release. Therefore, the length of time that a given mass of sludge is held in the stripper tank will affect the amount of acetate formed and thus phosphorus released. The rate of acetate formation per mass of sludge is also dependent on the overall activity of the sludge. An activated sludge operating at a high F/M will produce acetates more rapidly than a system at low F/M levels. Temperature also affects the rate of acetate formation which causes the release kinetics to exhibit an Arrhenius-type temperature relationship. Thus, the phosphorus level in the sludge leaving the stripper and entering the aeration basin can be controlled by the retention time in the stripper. But one has to keep in mind that excessively long retention does not necessarily increase the release rate because the total releasable phosphate is limited by the polyphosphate content of the organisms.

Therefore, to minimize stripper tankage volume requirement, a balance must be struck between retention time and throughput so that desirable P/VSS levels are achieved in the sludge. Furthermore, the "elutriation efficiency" in the stripper tank will have an effect since less phosphorus per unit mass needs to be released if a larger fraction of the released phosphorus is driven into the supernatant of the stripper tank.

TESTS AT RENO/SPARKS, NEVADA

After the preferred elutriation mode of operation was very successfully operated at Tonawanda, N.Y., the Environmental Protection Agency at Cincinnati agreed to partially fund a demonstration of the process at the Reno/Sparks wastewater treatment plant.

Pilot Plant Tests

The program consisted of two parts. One was carried out by means of a mobile PHOSTRIP pilot plant assembled by Union Carbide Corporation. The nominal feed rate to the van was 0.11 m^3/hr. (0.5 gpm). Primary settled wastewater from the full scale plant was used as feed.

One phase was operated such that a small portion of the influent to the pilot plant was used as elutriant in the stripper. An intensive evaluation period was carried out in late February and early March of 1976. The supernatant was allowed to go to drain.

Another phase was operated with lime precipitation such that phosphorus removal occurred only via the chemical and the waste biological sludges.

Table 4 gives a summary of the results, showing entirely satisfactory phosphorus removal with both operating modes.

Full Scale Testing

The plant is designed to handle an average flow of 76,000 m^3/d (20 mgd) and discharges into the Truckee River. It has three modules, each consisting of a square primary clarifier, a three-pass serpentine design aeration basin equipped with coarse bubble diffusers, and a secondary clarifier also square shaped. The combined primary and secondary waste sludges are anaerobically digested, dewatered on sand beds and trucked to a land fill.

The sludge recycle mode of operation had already been successfully tested in 1975 by L. E. Peirano[28] from Kennedy Engineers of San Francisco in June 1975 in cooperation with plant personnel. They had used one primary clarifier as stripper and were able to handle one third of the influent flow through the PHOSTRIP system.

Union Carbide in 1976 further modified the plant for full scale testing of the elutriation mode of operation. Again one of the primary clarifiers was converted to a stripper tank, while the two remaining primaries handled the entire flow. This time, however, two of the trains were used for the PHOSTRIP test, while the third was retained to operate in a conventional manner for comparison purposes. Since each train received the same amount of influent, the PHOSTRIP system handled two thirds of the influent.

Initially, some difficulties were encountered due to off-on type of operation of one of the feed pumps caused by the motor control center being inoperative. These difficulties were overcome by adjusting the flow rates around the stripper as a function of the high and low flow periods during the day. Normally this is not necessary. The operating conditions and results of the full scale tests are listed in

TABLE 4. AVERAGE ANALYTICAL RESULTS AT PHOSTRIP PILOT PLANT, RENO/SPARKS, NEVADA

	LPE (2/23-3/7/76)	LPE & Precipitation (3/20-4/2/76)
COD_{total} mg/1		
Influent	233	241
Effluent	26	19
Total Suspended Solids, mg/1		
Influent	93	98
Effluent	17	27
Effluent Volatile Suspended Solids, mg/1	14	21
Total Phosporus, mg P/1		
Influent	8.4	9.4
Effluent (ortho)	0.7 (0.5)	0.8 (0.5)
Stripper Supernatant	33	56
Stripper Underflow	216	386
Filtered Stripper Underflow	30	45
Filtered Aeration Effluent	0.8	0.8
Secondary Clarifier Underflow	190	338
Chemical Clarifier Overflow (ortho)	6.7 (3.5)	11.3 (2.4)
P/VSS		
Secondary Clarifier Underflow	0.042	0.048
Stripper Underflow	0.028	0.038

Tables 5 and 6, respectively.

As these tables show, the operation was entirely successful despite the difficulties with the large stepwise flow variations and the non-optimum arrangement caused by the need to convert the existing equipment to the PHOSTRIP system.

FURTHER PILOT PLANT TESTS

In addition to the tests conducted at Tonawanda and Reno/Sparks, several pilot plant programs were carried out for potential customers. These were done on location at Texas City, Texas; Adrian, Michigan; Brockton, Mass.; Findlay, Ohio; and at Tonawanda on imported waste from Southtowns, N. Y. All of these programs were fully successful as can be seen from the results listed in Table 7.

It is noteworthy that the operation at Texas City and at Adrian, Michigan were carried out by the plant operating personnel who found it easy to run the unit after a very short instruction period.

The programs at Adrian, Michigan and Findlay, Ohio are also somewhat unusual because in both instances it was found that the conventional process for phosphorus removal: namely, straight chemical precipitation - operated in parallel to the PHOSTRIP system - required unusually high quantities of chemicals. Also, the Findlay plant uses a contact-stabilization type flow scheme.

The operation using the Southtowns, N.Y. wastewater as influent resulted in a soluble phosphorus concentration of less than 0.1 mg/l in the effluent.

The test program at Brockton, Mass. was also highly successful, and has been reported on by Campbell et al.[29]

Summarizing, the process proved to be capable of removing the phosphorus from difficult-to-handle wastewaters using conventional flow schemes or a contact-stabilization (step feed) flow arrangement, at summer and at winter temperatures, and also proved to be easy to operate after a short familiarization period.

ECONOMICS

Many cost comparisons between the conventional precipitation type of phosphorus removal scheme and the PHOSTRIP system have been made. Because the processes are basically so different, the total yearly cost has been taken as the basis for comparison.

Several actual plant applications have been studied together with various hypothetical cases. Some of the results are listed in Table 8. It is clear from the data that the cost of the conventional precipitation process is essentially represented by the cost of the chemicals, while

TABLE 5. OPERATING CONDITIONS FOR FULL-SCALE TESTING AT RENO/SPARKS, NEVADA PHOSTRIP LPE SYSTEM (JANUARY 18-30, 1977)

Parameter	Minimum Train 1/Train 2	Maximum* Train 1/Train 2	Flow Weighted Mean Train 1/Train 2
Flow Rates, m^3/day			
Feed Q, (each train)	17280/17280	29380/29380	22460/22460
Recycle to Stripper	1987/2160	3200/3370	2510/2680
Stripper Underflow	1210/1470	2250/2250	1640/1810
Stripper Supernatant	6570	7000	6740
Elutriation	5270	5620	5440
Sludge Wasting	-	-	331/418
Aeration Time, Hrs. (based on Q)	9.0/9.0	5.1/5.1	6.8/6.8
Anoxic Period, Hrs. (based on R_3)	11.6	7.0	9.5
MLSS, mg/1	1230/1280	1110/1140	1170/1220
VSS/TSS	0.8	0.8	0.8
R_1SS, mg/1	5760/5460	6770/6720	6210/6030
VSS/TSS	0.78	0.78	0.78
TSS Stripper Supernatant, mg/1	-	-	59
VSS Stripper Supernatant, mg/1	-	-	54
pH, Mixed Liquor, mg/1	7.1/7.0	7.0/7.0	-
Temerature (Influent), °C	13	13	13
F/M_a(kg BOD_5/day/kg MLVSS)	0.45	0.88	0.62
Secondary Clarifier Overflow Rate, m/day	26.4/26.4	44.8/44.8	34.3/34.3

TABLE 6 ANALYTICAL RESULTS FOR FULL-SCALE TESTING AT RENO/SPARKS, NEVADA PHOSTRIP LPE SYSTEM (JANUARY 18-30, 1977)

Parameter	Train 1/Train 2	Train 1/Train 2	Flow Weighted Mean Train 1/Train 2
BOD_5, mg/l			
Influent	-	-	168
Effluent	-	-	28/29
Total Suspended Solids, mg/l			
Influent	-	-	102
Effluent	-	-	21/23
Effluent Volatile Suspended Solids, mg/l	-	-	∿18
Total Phosphorus, mg P/l			
Influent	-	-	9.1
Effluent (ortho)	-	-	0.8(.5)/0.9(.6)
Stripper Supernatant	-	-	36
Stripper Underflow	255	270	260
Filtered Stripper Underflow	50	70	59
Secondary Clarifier Underflow	215/195	280/250	245/220
P/VSS			
Secondary Clarifier Underflow	.048/.046	.053/.048	.05/.047
Stripper Underflow	0.028	0.030	0.029

TABLE 7. RESULTS OF PILOT-SCALE DEMONSTRATION PROGRAMS

Location	Texas City, TX	Adrian, MI	Southtowns, NY	Brockton, MA	Findlay, OH
Duration of Phase, days	14	31	19	26	12
Feed Rate (Q), l/s	.011	.009	.0083	.24	.0073
Recycle Rate, % of Q	14	20	17	37	30
Stripper Feed Rate, % of Q	14	17	7	20	22
Stripper Return to Aeration, % of Q	12	15	7	18	20
Stripper Supernatant, % of Q	17	17	1.3	14	20
Elutriation Rate, % of Q	15	15	1.3	12	20
Influent Characteristics, mg/l:					
BOD_5 (COD)	73	61	121	212	(315)
Total Phosphorus	7.9	8.4	3.0	11.3	5.9
Total Suspended Solids	69	74	89	344	62
Temperature (Aeration), °C	22.7	26.7	18.9	12.2	13.0
Total Suspended Solids (Mixed Liquor), mg/l	1,672	2,244	2,980	2,525	2,000
F/M (BOD_5/day/kg MLVSS)	0.4	0.3	0.4	0.5	0.6 (COD basis)
Effluent Characteristics, mg/l:					
BOD_5 (COD)	29	16	34	11	(45)
Total Effluent Suspended Solids	25	35	51	15	11
Total Phosphorus	0.7	1.6	1.4	0.8	0.7
Ortho Phosphorus	0.3	0.2	<0.1	0.3	0.1

TABLE 8. COST OF PHOSPHORUS REMOVAL SYSTEM

Location	Hypothetical		Brockton, Mass.		Reno, Nev.		Hypothetical	
	Conv.	PhoStrip	Conv.	PhoStrip	Conv.	PhoStrip	Conv.	PhoStrip
Plant Size, mgd	10		18		40		100	
Installed Investment ($000)	31	524	150	1150	84	2084	94	4084
Annual Cost, $000/yr:								
Amortization[1]	2.7	46.2	13.2	101.3	7.4	183.6	8.3	359.7
Chemical: Alum[2]	155.3	-	246.6	-	745.	-	1553.	-
Lime[3]	-	24.0	24.7	41.1	-	144.2	-	240.1
Operation, Maintenance and Repair	5	15	7.3	17	10	22	25	45
Totals	163.0	85.2	291.8	159.4	762.4	349.8	1586.3	644.8
Specific Costs, $/MG	44.7	23.2	44.4	24.3	52.2	24.0	43.5	17.7

(1) 20 year project life, 6-1/8% interest.
(2) $80/t delivered.
(3) $40/t delivered.

the PHOSTRIP system costs are dominated by the capital costs. Furthermore, the total yearly cost of the PHOSTRIP system is about half of that of the conventional chemical precipitation process, the difference being somewhat less at the small sizes and somewhat more at the largest size studied. Of course, for specific situations, the actual costs can change somewhat because of different amounts of chemicals required, different costs of chemicals delivered, and varying economic considerations.

Cost comparisons for specific installations have also been made by several Consulting Engineers. One comparison was released by McNamee, Porter and Seeley from Ann Arbor, Michigan, who have studied phosphorus removal for the planned wastewater treatment plant for Adrian, Michigan, which has a capacity of 7 mgd. The cost estimate for the PHOSTRIP system, including sludge disposal costs, came to 62.8% of that for the alum precipitation process, which finding the Engineers used to request approval for a sole source specification of the PHOSTRIP system for Adrian.

SIMULTANEOUS PHOSPHORUS AND NITROGEN REMOVAL

The second most important nutrient which needs to be minimized in wastewater effluents is nitrogen. Conventional activated sludge processes discharge the soluble nitrogen in the form of ammonia. This ammonia exerts a considerable amount of BOD on the receiving waters, and whenever this is critical the plant has to be designed for nitrification. Nitrification is the biological oxidation of ammonia to nitrate. This is a two-step process where the ammonia is first oxidized to NO_2 by *Nitrosomonas* species and then to NO_3 by *Nitrobacter*. The nitrification process is well known and need not be described here. Suffice it to say that the key to the process design is to ensure that the sludge retention time is long enough so that the nitrification bacteria which grow much slower than the carbonaceous bacteria are not washed out but maintained at a sufficiently high concentration to be able to accomplish nitrification.

Nitrification, although it removes the BOD exerted by ammonia, does not remove the nitrogen from the water, and, in fact, excessive nitrate concentration has some undesirable characteristics.[30] If the nitrate in the effluent cannot be tolerated, then one has to resort to denitrification. This is an additional biological step, operating primarily under anoxic conditions where some of the bacteria utilize the nitrate as source of oxygen.

Early process designs for nitrification consisted of a one- or two-stage step for carbonaceous removal and nitrification and a subsequent anaerobic step for denitrification. This latter step is designed for using methanol as carbon source. Other designs[30] are based on having alternate aerobic and anaerobic stages on the main stream.

Since the PHOSTRIP system already has an anoxic tank, it is relatively easy to extend the process to denitrification. The aerobic portion has to be laid out, of course, for nitrification. As mentioned above, this can be achieved by designing for a sufficiently long sludge retention time. The flow diagram of the combined PHOSTRIP system-denitrification process is shown in Figure 7. The underflow from the clarifier is conducted into a covered tank without aeration where the sludge is exposed to anoxic conditions for about two hours and undergoes denitrification. Nitrogen gas is discharged at the top. The effluent sludge then is split; about half goes back to the aeration tank, and the other half is piped into the stripper tank. There, as before, the sludge since exposed to continued anoxic conditions releases orthophosphate. This is driven by elutriation into the supernatant from where it is removed by precipitation. The underflow from the stripper containing sludge partially depleted of phosphorus is returned to the aeration basin where, exposed to aerobic conditions, it again takes up the phosphorus brought in by the influent wastewater.

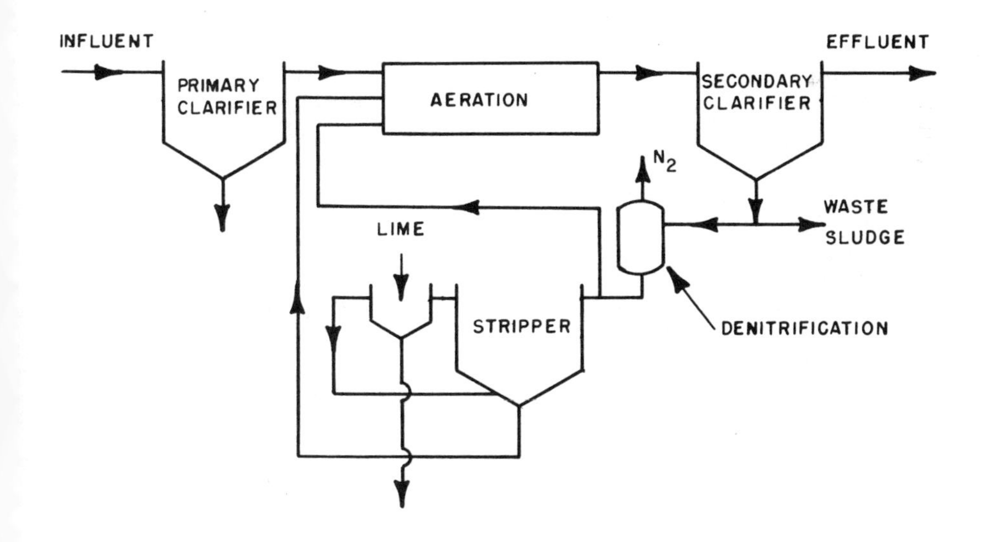

FIGURE 7. COMBINED PHOSPHORUS-NITROGEN REMOVAL

If the clarifier underflow rate is 30% of the influent, one would initially assume that the denitrification rate is about 0.3/1.3 or 23%. This is, however, not the case at all. In pilot plant operations at Tonawanda, N.Y., this process was tested on municipal wastewater, and the overall denitrification rate observed was about 70%. The only explanation which can be offered for this unexpectedly high rate of denitrification is that a considerable amount of denitrification occurs in the aeration basin. It appears that in this system some bacteria continue to utilize nitrate as ultimate electron acceptor for a period of time after entering the aeration basin. Organisms capable of utilizing nitrate have been reported in the literature.[31]

In summary, the combination of the PHOSTRIP system with a denitrification tank in the sludge return line is a very simple one-stage system. It is able to remove the phosphorus in the effluent to very low levels, to remove about 70% of the nitrogen, and to discharge the remaining nitrogen in the form of nitrate. The simplicity of the process and the fact that it uses very little lime for phosphorus removal and no methanol but impurities in the influent as source of carbon for denitrification makes it economically most attractive.

REFERENCES

1. Process Design Manual for Phosphorus Removal. 1976. U. S. Environmental Protection Agency Technology Transfer.

2. Ernst, P. 1888. Z. Hyg. IV, 25.

3. Babes, V. 1889. Z. Hyg. V, 173.

4. Grimme, A. 1902. Cbl. Bakt., Abt. I, Orig., XXXII,191.

5. Meyer, A. 1904. Bot. Z., Abt. I, LXII, 113.

6. Zikes, H. 1922. Cbl. Bakt., Abt. II, LVII, 21.

7. Duguid, H. P. 1948. J. of Pathology and Bacteriology, LX, 265.

8. Kornberg, A., S. R. Kornberg, and E. S. Simms. 1956. Metaphosphate Synthesis by an Enzyme from Escherichia Coli. Biochem. Biophys. Acta 26:215-227.

9. Srinath, E. G., C. A. Sastry, and S. C. Pillai, 1959. Rapid Removal of Phosphorus from Sewage by Activated Sludge. Experientia 15, 339.

10. Feng, T. H. 1962. Phosphorus and the Activated Sludge Process. Water and Sewage Works. 109, 431.

11. Sekikawa, Y., S. Nishikawa, M. Okazaki, and K. Kato. 1966. Release of Soluble Orthophosphate in Activated Sludge Process. J. WPCF. 38, 3, 364.

12. Levin, G. V. 1966. U. S. Patent No. 3,236,766.

13. Levin, G. V., G. J. Topol, and A. G. Tarnay. 1975. Operation of Full-Scale Biological Phosphorus Removal Plant, J. WPCF. 47, 3, 577.

14. Fuhs, G. W. and Min Chen, 1974. Microbiological Basis of Phosphate Removal in the Activated Sludge Process for the Treatment of Wastewater. Presented at the XIX Congress of the International Society for Theoretical and Applied Limnology, Winnepeg, Canada.

15. Russ, C. 1975. Microbial Enhancement of Phosphorus Removal in Sludge Sewage Systems. PhD Dissertation, Microbiology Department, University of Arizona.

16. Kornberg, S. R. 1957. Adenosine Triphosphate Synthesis from Polyphosphate by an Enzyme from Escherichia Coli. Biochimia et Biophysica Acta. 26:294.

17. Butler, L. 1977. A Suggested Approach to ATP Regeneration for Enzyme Technology Applications. Biotechnology Applications. Biotechnology and Bioengineering 19:591.

18. Fox, J. L. 1977. Pyrophosphate Drives Biochemical Reactions. Chemical and Engineering News, 22 (April 25).

19. Heffley, P.D. 1977. Phosphate Release Studies. Memorandum to R. F. Drnevich, Union Carbide Corporation internal communication.

20. Russ, C. 1977. Personal communication. The Pennsylvania State University.

21. Gould, M. S. 1976. Personal communication. Union Carbide Corporation.

22. Levin, G. V. 1974. Personal communication. Biospherics, Inc.

23. Sheridan, D. 1977. Personal communication. The Pennsylvania State University.

24. Matsch, L. C. and R. F. Drnevich. 1977. U. S. Patent No. 4,042,493.

25. Drnevich, R. F. and L. M. LaClair, 1976. New System Cuts Phosphorus for Less Cost. Water and Wastes Engineering. 104.

26. Dick, R. I. 1970. Role of Activated Sludge Final Settling Tanks. Journal Sanitary Engineering Division, Amiercan Society of Civil Engineers. 96, SA2: 423.

27. Dick, R. I. and K. W. Young. 1972. Analysis of Thickening Performance of Final Settling Tanks. 27th Annual Purdue Industrial Waste Conference. Lafayette, Indiana.

28. Peirano, L. E. 1977. Low Cost Phosphorus Removal at Reno/Sparks, Nevada. J. WPCF, 49, 4; 568.

29. Campbell, T. L., J. M. Reece, and T. J. Murphy. 1977. Presented at the New England Water Pollution Control Association Annual Meeting. Whitefield, N.H., June 13-14.

30. Barnard, J. L. Water Pollution Control 1973, p 705.

31. Bergeys Manual of Determinative Bacteriology. 8th edition, Part 7, p 217.

8. HYACINTHS

James S. Taylor. Civil Engineering and Environmental Sciences Department, Florida Technological University, Orlando, Florida

E. A. Stewart. Dawkins and Associates, Inc., Orlando, Florida

ABSTRACT

This paper contains a review of existing literature that is pertinent to the utilization of water hyacinths for wastewater treatment by maximizing plant productivity. New productivity of water is critical to cost effective hyacinths wastewater treatment and varies from 0.45 to 16 ton/ac-day. Optimum productivity occurs at pH 7 and 27°C in an aerobic environment that provides adequate nutrients and space for new growth. Plant harvesting is required to prevent a nutrient release from sediment material that will accrue in a natural environment.

System design parameters of 23.3 lbs BOD/ac-day, 20.4 lbs SS/ac-day, 5.4 lbs N/ac-day and 1.6 lbs P/ac-day were developed from existing literature on plant productivity. These parameters were used in the synthesis of 17 equations that provide a cost estimate of three water hyacinth systems utilizing compost, cattle feed, and landfills as means to dispose of the harvested plants. The cost analysis indicates that a market must be developed for water hyacinth systems to be cost effective relative to landspreading or AWT.

INTRODUCTION

Water hyacinths (Eichhornia crassipes) are highly productive aquatic plants that could possibly provide a cost effective method of nutrient removal. Any treatment system which utilizes water hyacinths will have to manage several parameters which are vital to the success of the system. Previous investigations into water hyacinth nutrient removal systems have been conducted but none of these has attempted to optimize nutrient removal through chemical and physical control.

This paper (1) describes the factors that are critical to nutrient removal by water hyacinths, (2) reviews the previous systems that have employed water hyacinths for nutrient removal, (3) relates these factors to optimizing a nutrient removal system utilizing water hyacinths, and (4) estimates the cost as a function of flow for differing conceptional systems which utilize water hyacinths for waste treatment.

UNITS

Several different growth parameters and units have been used to describe the water hyacinth. Examples of the most typical parameters are growth rate, doubling time, geometric factors, productivity, biomass, crop, standing crop and yield. Since the purpose of this paper is to evaluate water hyacinths on a comparative basis with other nutrient removal alternatives, and the mechanism is assumed to be plant uptake, net productivity is the most critical parameter. Where possible, all seasonal plant productivity will be reported in lb/ac/time with the metric units of kg/m^2/time given in parenthesis. Annual productivity will be reported in units of tons/hectare/year or lb/ha/yr with mt/ha/yr given in parenthesis. Daily productivity will be reported as lb/ac/day with gm/m^2/day given in parenthesis. These productivity units will refer to both wet and dry weights as specified, and are derived from direct weighings rather than measurement of photosynthetic activity. Any reference to exponential growth rates will be made on a daily percentage basis and will refer to wet weight only. Gross productivity will be given in tons/ac or mt/ha which will only serve as an estimate of surface density and should not be misconstrued as an estimate of potential nutrient removal.

COMPOSITION

Boyd and Blackburn[5] harvested water hyacinths from natural plant infestations in Ft. Lauderdale, Florida from April through August on a monthly basis. Their results demonstrate that the percent dry matter of the aquatic plant and the crude protein, ether extract and cellulous content of the hyacinth varies significantly by season. Their results are presented in Table 1 and the percent nitrogen variability has been determined stoichiometrically from this data. It should be noted that the percent nitrogen content of the plant coupled with growth rate are crucial to any nutrient removal alternative proposing use of water hyacinths.

Parra and Hortenstein[22] collected water hyacinth samples from 19 different natural locations through the state of

TABLE 1.

MONTHLY CHANGES IN COMPOSITION OF WATER HYACINTHS TAKEN FROM A NATURAL LOCATION IN FORT LAUDERDALE, FLORIDA

Month	% Dry Matter	% Dry Weight		
		Crude Protein	Ether Extract	Cellulose
April	5.0	22.0	5.29	25.7
May	5.0	23.5	5.60	26.7
June	8.0	18.2	3.75	22.8
July	7.3	15.7	5.11	21.5
August	7.0	19.4	5.84	20.4

Florida. The results of this collection are presented in Table 2 again as a percentage of plant dry weight. The ratio of C/K/N/P based on percent P in the dry weight of the water hyacinth is 113/12.3/5.2/1. The water content of the fresh plants was 95%. The average dry weight composition of the water hyacinths was 19.2% ash, 34.9% C, 1.61% N, 0.31% P, 3.81% K, 1.66% Ca, 0.56% Mg, 0.56% Na, 2568 ppm Al, 2772 ppm Fe, 286 ppm Mn, 58 ppm Zn and 9 ppm Cu. This is by far the most complete sampling of water hyacinth composition found by the authors although the nitrogen content is less than many of the values that were reported by other investigators.

Boyd[6] reported water hyacinth composition based on data obtained by sampling plants from only young stands. The K/N/P ratio of their data reported is 9.9/6.2/1. This ratio falls within the ranges experienced in the Parra and Hortenstein[22] study reported earlier. The nitrogen concentrations reported are within the nitrogen values that were determined from the Boyd and Blackburn[5] study which could indicate that if the differences in plant composition are due to the emergence of young plants, these young plants emerge at different periods within the same year.

The main constituent of water hyacinths is water. Authors have reported water compositions in water hyacinths varying from 94 to 96%. These studies have not been done on a seasonal basis and variations of 92 to 95% were reported by Boyd[5]. Penfound and Earle[23] analyzed the water contained in ten water hyacinths from the New Orleans area and found an average concentration of 96.05%. Significant variability (from 89.3% in the blades to 96.7% in the stolens) was noted in different parts of the plants.

TABLE 2.

CHEMICAL COMPOSITION OF WATER HYACINTH TAKEN FROM SEVERAL NATURAL FLORIDA LOCATIONS
(Composition Reported as Percent Dry Weight)

Origin	Ash	C	N	C/N ratio	P	K	Ca	Mg	Na
Lake Istokpoga (Sebring)	24.4	18.0	1.08	16.7	0.14	1.00	0.73	0.38	0.15
Lake Eden Canal (SR 532)	19.4	28.8	0.86	33.5	0.09	1.95	0.46	0.31	0.23
Lake Thonotosassa	23.0	23.0	1.17	19.7	0.33	3.35	1.49	0.29	0.21
Waverly Creek (SR 60)	25.0	33.1	2.26	14.6	0.56	3.10	1.58	0.50	0.37
Arbuckle Creek	23.4	34.9	1.90	18.4	0.23	3.35	1.06	0.49	0.28
Lake Tohopekaliga (Kissimmee)	21.7	34.0	1.69	20.1	0.60	4.70	1.56	0.71	0.53
Lake Monroe (Sanford)	20.4	32.5	2.86	11.4	0.59	5.55	1.73	0.54	0.83
Duda Canal No. 1 (Belle Glade)	20.3	39.1	1.30	30.1	0.13	3.80	1.99	0.60	0.48
St. Johns River (Astor)	20.1	36.4	2.33	15.6	0.51	6.50	1.43	0.51	0.63
W. R. Grace Landfill (Bartow)	19.0	36.4	1.86	19.6	0.59	2.72	1.99	0.56	1.54
Ponce de Leon Springs	18.5	37.5	1.74	21.5	0.33	5.40	2.34	0.50	0.47
Waverly Creek (SR 540)	18.5	38.1	1.76	21.6	0.32	4.85	1.45	0.55	0.67
Duda Canal No. 2 (Belle Glade)	17.5	37.8	1.66	22.8	0.15	4.70	2.28	0.69	0.57
Lake Alive (N. of Florida)	17.3	38.6	1.17	33.0	0.40	3.66	2.41	0.69	0.40
Lake Apopka (Monteverde I)	15.8	38.8	1.22	31.8	0.14	4.26	2.07	0.54	0.41
St. Johns River (Palatka)	15.8	38.0	1.82	20.9	0.16	3.44	1.83	0.73	0.86
Lake George	15.4	40.2	1.48	27.1	0.21	3.21	1.91	1.86	1.24
Lake Apopka (Monteverde II)	14.9	39.8	1.36	29.3	0.09	4.08	1.96	0.60	0.21
Lake East Tohopekaliga (St. Cloud)	14.7	37.2	1.08	34.5	0.23	2.90	1.19	0.51	0.53
Mean	19.2	34.9	1.61	23.3	0.31	3.81	1.66	0.56	0.56
Standard Deviaticn	3.2	5.9	0.50	7.0	0.18	1.30	0.53	0.14	0.36

Source: Parra, J.V. and Hortenstein, C.C. Plant Nutritional Content of Some Florida Water Hyacinths and Response by Pearl Mullet to Incorporation in Three Soil Types. Hyacinth Control Journal, presently known as The Journal of Aquatic Plant Management. Reprinted with Permission. Copyright 1974, 12: 85-90.

PRODUCTIVITY

Growth Rate

Many investigators have reported on water hyacinth productivity on a gross, seasonal and daily basis measuring surface area, total mass or partial mass. Investigators at the University of Florida under the direction of Prof. T des Furman[9] reported a surface coverage rate of 11.2% during April and May in a lagoon fed by secondarily treated sewage effluent. This growth rate is approximately twice the maximum growth rate of water hyacinths in natural environments and it is noted that during these spring studies there was always an abundance of space and nutrients in the lagoon. However, limited density measurements indicated that the surface density during this growth phase was approximately half, 681 tons/ac (250 mt/ha), of that reported by Wahlquist[32]. Water hyacinths were grown in 0.04 ha by Wahlquist[32] to determine their response to various N/P/K combinations. Experiments indicated that only nitrogen would greatly affect the yield of water hyacinths. Standing crops as wet weight of water hyacinths were reported as 474.64 tons/ac (174.5 mt/ha) from 0/0/0 treatment, 1497.1 tons/ac (550.4 mt/ha) from 0/8/0 treatment, and 1607.3 tons/ac (590.9 mt/ha) from 0/8/8 treatment. Obviously, the necessary amounts of potassium and phosphorus significant enough to produce a large hyacinth yield were obtainable from the soil or other sources. This study was conducted from April through November which included the peak water hyacinth growing season.

Yount and Crossman[38] reported a hyacinth growth rate of 4.08% to 1.78% through the warmer winter months in Winter Haven, Florida. Although these studies were conducted on a batch basis, nutrients were added and harvesting was performed periodically to provide space for new growth. Penfound[23] reported that doubling times based on the number of new plants produced on the field near New Orleans, LA. were 4.6%, 5.5%, 6.2% per day on the basis of observations of three different sites.

The most complete report of water hyacinth growth rates was reported by Bock[4] in her doctoral dissertation at the University of California at Berkely. Bock[4] conducted three field studies of water hyacinth growth in a river near San Joaquin, California. In the first field study, although no harvesting was conducted, several plants were lost due to an impoundment breakdown. The growth rates from Bock's work are reported as percentage of mass increase based on wet weight in Table 3.

Although complete yearly data was not obtained by Bock, the significance of adequate space requirements for optimum water hyacinth mass-yield is highlighted by her work. The highest daily growth rates in excess of 5% were typically

TABLE 3.

PERCENT OF WATER HYACINTH MASS INCREASE BASED ON WET WEIGHT FROM THREE STUDIES PERFORMED IN A RIVER NEAR SAN JOAQUIN, CAL.

Study	1		2		3
Month	% Growth	Month	% Growth		% Growth
August	5.41	April	5.67	April	5.00
September	2.42	May	4.20	May	2.64
October	1.20	June	5.60	June	4.28
November	1.22			July	3.93
				August	1.78
				September	1.06
				October	-0.29

Source: Bock, J.H. An Ecological Study of Eichhornia Crassipes with Special Emphasis on its Reproductive Biology, Ph.D. Dissertation, 1966, University of California, Berkeley, CA.

found at the beginning of each study when space was plentiful. Further comparison of the August growth rate in the first and third study shows significantly different daily growth rates of 5.41 and 1.78%. The latter occurring in the fifth month of an eight-month study when little free space was available.

The typically exponential growth of water hyacinths in the natural or man-made environment can be made evident by presenting data from a Wolverton and McDonald[34] study in a semi-log plot as shown in Figure 1. Exponential growth of water hyacinths was observed throughout all periods of the study. The data was fitted to the exponential equation 1 by the authors with a resulting correlation coefficient of 0.97. Even though the correlation coefficient is high for the semi-logarithmic equation, significant deviations of actual from predicted surface area is experienced in June which emphasizes the need for seasonal rates to accurately predict water hyacinth growth.

$$SA = e^{0.0307t + 8.3} \tag{1}$$

SA = Surface area in ft^2
t = time of growth in days

Given: Original SA approximates 0.05 HA, growth begins in mid-March ($\overline{T} = 16.3^{o}C$) and continues to mid-June ($\overline{T} = 26.3^{o}C$) in Mississippi, no growth space restrictions, plants grown in raw sewage.

As can be seen in Figure 1, the largest growths occurred with the highest monthly average ambient temperature. The growth rates have been presented in Table 4. The data infers that the growth rate is a nonlinear function of temperature with 26.3°C being the optimum average ambient temperature of those investigated. Limited data is available on the effect of temperature on water hyacinth growth. Most investigators have simply reported that hyacinth growth does not occur and death results from prolonged exposures to temperatures lower than 32°F.

TABLE 4.

GROWTH RATE OF WATER HYACINTHS CORRESPONDING TO MONTHLY AVERAGE AMBIENT TEMPERATURES

Month	Time (Days)	Area (HA)	Average Ambient Temp °C	Rate %
March	10	0.04	16.0	-
April	40	0.12	18.7	3.66
May	71	0.33	22.9	3.26
June	89	0.88	26.3	6.54

Source: Wolverton, B.C. and R.C. McDonald. (Oct. 1976) Water Hyacinths for Upgrading Lagoons to Meet Advanced Wastewater Treatment Standards: Part II. TMX-72730, NASA, NSTL, Bay St. Louis, Miss.

However, Bock[4] published data on growth chamber studies that indicate optimum water hyacinth growth occurs at a temperature of 26.7°C. The results of her growth chamber studies are shown in Figure 2. Although Bock's studies were intended to emphasize the effect of photoperiod on water hyacinth growth, they also demonstrate that temperature is very significant in regards to productivity of the water hyacinths. The results of Boyd[4] and Wolverton and McDonald[34] also indicate that as temperature increases, the rate of hyacinth growth will also increase. Penfound and Earle[23] reported that water hyacinths could not survive water temperatures of 34°C for more than five hours. They investigated the effect of freezing temperatures on water hyacinths as a function of time. The results of their investigations are given in Table 5.

The authors were unable to find any studies which determined the optimum temperature for hyacinth growth and insured adequate space and nutrients for growth during the study. However, it may be possible to use a heated waste effluent in combination with a hyacinth treatment system to enhance nutrient removal.

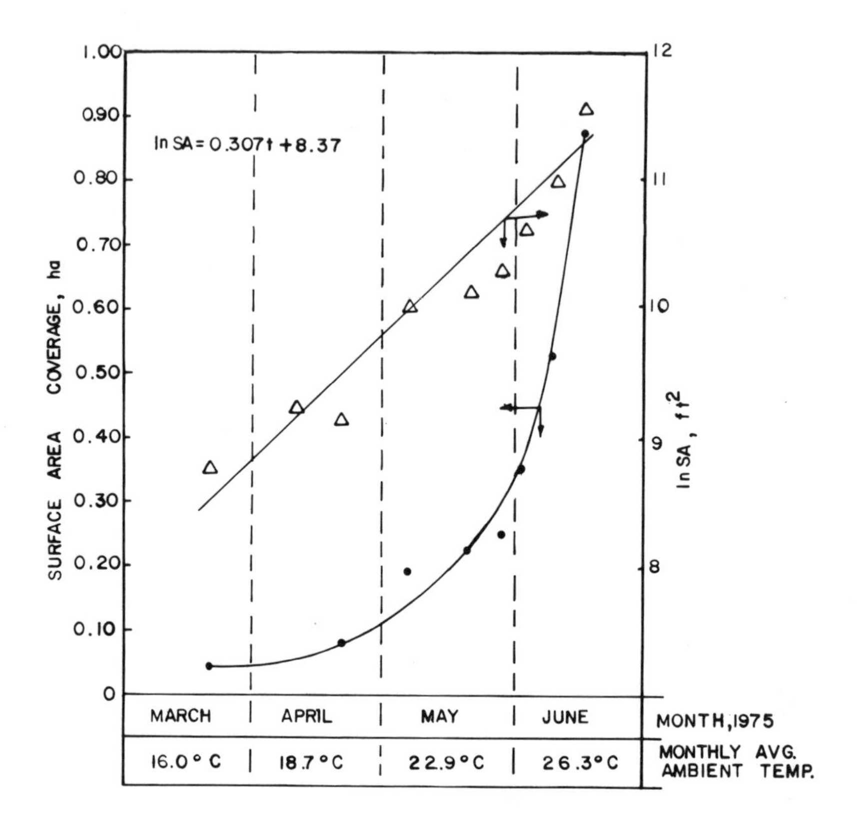

FIGURE 1. Surface Area Coverage of a Lagoon Receiving Raw Sewage by Water Hyacinths Corresponding to Time and Average Ambient Temperature

REFERENCE: Wolverton, B.C. and McDonald, R.C. "Water Hyacinths for Upgrading Sewage Lagoons to Meet Advanced Wastewater Treatment Standards", Part II, TM-X-72730, NASA, NSTL, Bay St. Louis, MS, October 1976.

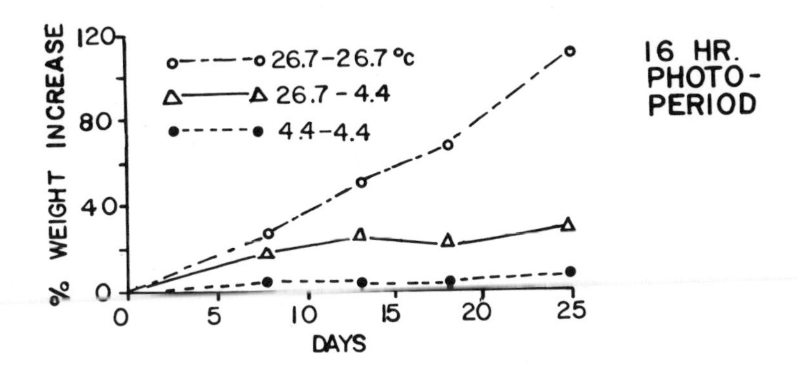

FIGURE 2. Percent of Water Hyacinth Weight Increase Corresponding to Growth Period and Temperature for an 8 Hour and a 16 Hour Photoperiod

REFERENCE: Bock, J.H. An Ecological Study of Eichhornia Crasspies with Special Emphasis on its Reproductive Biology, Ph.D. Dissertation, 1966, University of California, Berkeley, CA.

TABLE 5.

EFFECT OF FREEZING TEMPERATURES ON WATER HYACINTHS

Temp (^{o}F)	Hours Exposed					
	Plant Injury			Resprouting		
	12	24	48	12	24	48
33	Blades	Blades	Blades	All	All	All
27	Blades Floats	Leaves Killed	Leaves Killes	All	All	All
23	Leaves	Leaves Killed	Dead	All	All	0
21	Leaves	Dead	Dead	0	0	0
19	Dead	Dead	Dead	0	0	0

Source: Penfound, W.T. and T.T. Earle. (1948) The Biology of the Water Hyacinth. Ecol. Mono. 18:447-478.

Mass

Several authors have reported productivity as crop yield in units of weight/surface area of water hyacinth/time, i.e., tons/ac/yr (mt/ha/yr) for annual productivity. Scarsbrook and Davis[27] studied the effects of sewage effluent on the growth of water hyacinth, Alligatorweed, Curly Pondweed,

Egeria and Slender Naiad which are vascular plants. Their studies were conducted on a batch basis (during two 11-week phases, April through October) in twelve water hyacinth containing plastic pools which were filled with sewage effluents. The results from the static system indicated that the first 11-week growth period returned a higher water hyacinth yield than did the second 11-week growth period. The nutrient uptakes are shown in Table 6. Scarsbrook and Davis [27] demonstrated three significant items regarding water hyacinths. First, they showed that of the vascular plants investigated, water hyacinths were by far the most productive. Secondly, they showed that water hyacinth productivity could be greatly increased by the nutrients present in treated sewage and finally, that the growth of the water hyacinth once stimulated by sewage effluent was greatly reduced as the nutrients were removed. Of the total increase in water hyacinth mass, 36.6 gm, only 14%, occurred in the second half of the study. It should be noted that at no point was space limited for water hyacinth growth in this work.

TABLE 6.

NITROGEN AND PHOSPHORUS UPTAKES BY WATER HYACINTHS IN A STATIC SYSTEM INITIALLY CONTAINING 25% SEWAGE EFFLUENT

Time (Days)	Mass in Grams					
	Water Hyacinth Dry Weight	Nitrogen	Phosphorus	N/P	%P	%N
April	2.0					
July	633.3	5.889	2.723	2.16	0.43	0.93
August	736.6	7.542	2.950	2.56	0.40	1.02

REFERENCE: Scarsbrook, E. and Davis, D.E. Effect of Sewage Effluent on Growth of Five Vascular Aquatic Species. Hyacinth Control Journal, Presently known as The Journal of Aquatic Plant Management. Reprinted with Permission. Copyright 1971, 13: 56-58.

A second investigation of water hyacinth productivity was conducted by Ornes and Sutton [21] in Fort Lauderdale, Florida. Water hyacinths were grown in static sewage effluent from May through July. Harvesting was conducted at five-week intervals to provide 50% free surface area for additional growth. The maximum dry weight yield, 22.6 lbs/ac/day (13.7 gm/m^2/day) occurred during the first week of the study during the time when nutrient concentrations were highest. As harvesting continued, the growth rate decreased as the available nutrients decreased. This study, as does Wahlquist [32] and Scarsbrook and Davis[27], demonstrates that water hyacinth productivity is directly dependent on nutrient concentration of the containing waters. The productivity of water hyacinths, and consequently the nutrient removal from the containing wastewaters, decrease as the concentration decreases. This phenomena will significantly influence the costs associated with removing nutrients from wastewaters to AWT standards. Somewhat surprising is the Ornes and Sutton[21] finding that as harvesting continued the pH increased from 6.9 to 8.0 during the study. Seemingly, pond coverage could reduce photosynthetic activity and allow increased CO_2 buildup in the containing waters.

Only two studies were found that involved data collection of production on a yearly basis. The first of these was conducted by Penfound and Earle[23] who determined the plant material present in a mat containing medium sized water hyacinths located in nearby New Orleans, LA. This data is presented in Figure 3 and does demonstrate the cyclic nature of water hyacinth production. Growth is seen to begin in March and continue to July. The one sample point in October is seemingly high and out of place when compared to the natural cycle of an increase in mass from March through July and a decrease in mass from August to November. The data would then suggest the plants existed at steady state during the winter months with constant growth and death rates at the New Orleans site. The method of sampling consisted of averaging the mass of 3 samples each of which had a surface area of 0.55 ft^2 and was meant to only approximate productivity. However, these values do fall within the range 680-1360 tons/ac/yr (250-500 mt/ha) of water hyacinth production as reported by other investigators. The maximum and minimum production values reported by Penfound and Earle[23] ranged from 410 to 210 mt/ha. It must be noted that the nutrient availability is unknown, harvesting was not performed and growing space restrictions almost certainly occurred in a natural environment.

A second productivity study was conducted by Yount and Crossman[38] in Winter Haven, FL. This investigation spanned three years and involved plant harvesting on a regular basis so as to provide 50% free space for plant growth. The results of the Yount and Crossman[38] study represents net productivity which would be more applicable to nutrient removal since it represents the rate of nutrient removal by water hyacinths. The study was conducted in ten vinyl swimming pools, each of which was initially dosed with 4 lbs of improved Milorganite and 20 grams of lime so that adequate nitrogen, phosphorus, calcium and iron was available to the water hyacinths for growth. In July, four pounds of water hyacinths were placed in metal containing rings within the pool which had a surface area of one square meter. The plants were allowed to grow within the containing rings until the square meter of surface area was exhausted and at this point plants from 50% of the surface area were harvested.

The harvested plants were weighed to determine productivity. Four of the ten pools were classified as control pools. All plants that were harvested from these pools were dried, ground, and returned to the pool in order to stimulate a natural stand of water hyacinths. The harvested water hyacinths from the six remaining ponds were weighed and discarded. Additional nutrients in the form of Milorganite were added to two of these six ponds which were classified as nutrient addition ponds. The four final ponds were described as test ponds. The results of this productivity investigation are shown in Figure 4.

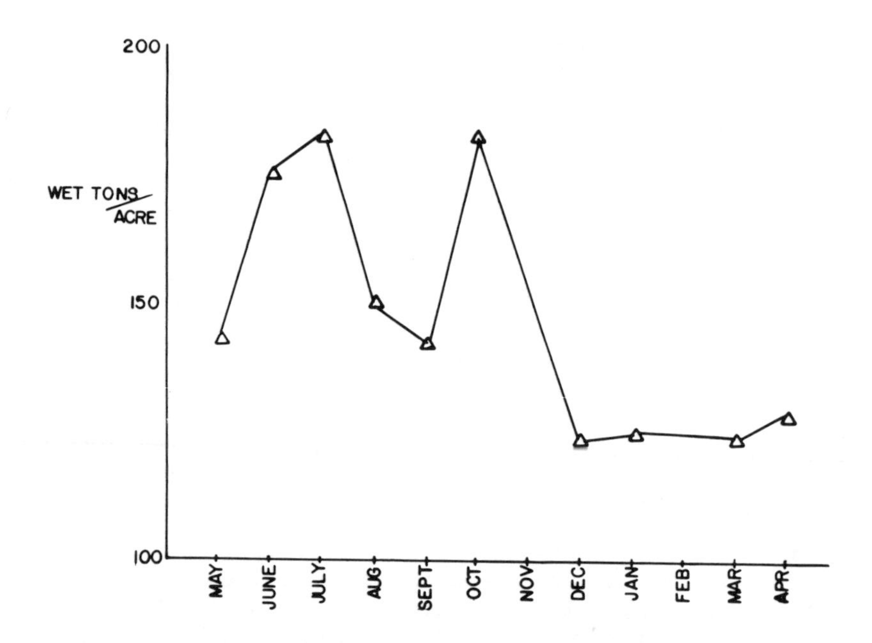

FIGURE 3. Natural Variation of Surface Density in a Mat of Unharvested Water Hyacinths Located Near New Orleans, Louisiana by Month.

REFERENCE: Penfound, W.T. and Earle, T.T. "The Biology of the Water Hyacinth", Ecological Monographs. Reprinted with Permission. Copyright 1948, Vol. 18, No. 4, by the Ecological Society of America.

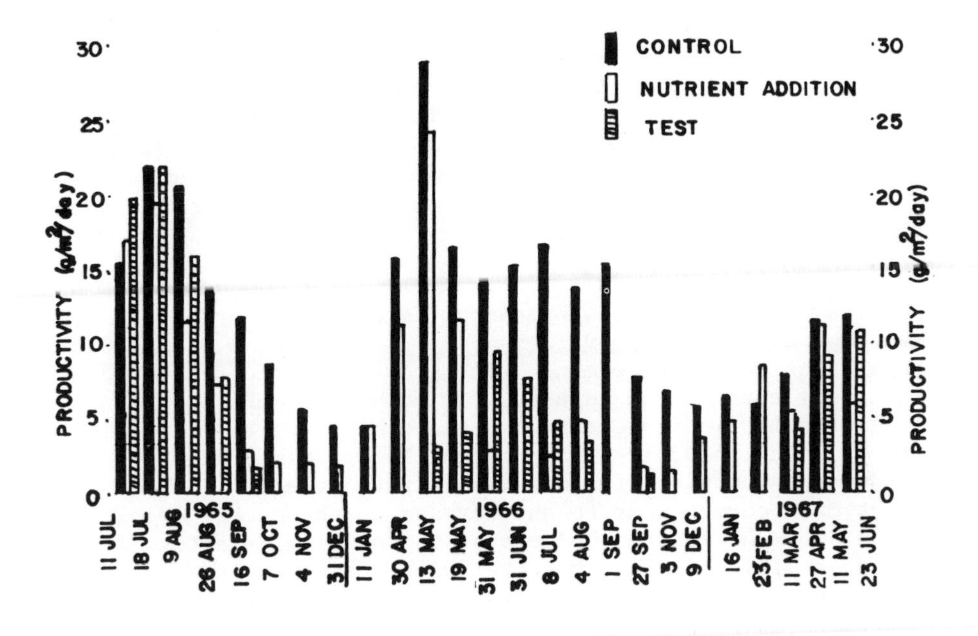

FIGURE 4. Average Dry Weight Productivity of Water Hyacinths in Test, Control and Nutrient Addition Experiments

REFERENCE: Yount, J.L. and Crossman, R.A., Jr. "Eutrophication Control by Plant Harvesting". Copyright Vol. 42, 1970, WPCF and reprinted with permission.

Yount and Crossman [38] demonstrated that the high productivity of water hyacinths would be continued on a seasonal basis if the nutrients removed by the plants were allowed to return to the water. Productivity ranged from 259 lb/ac/day (29 $gm/m^2/day$) in May 1966 to 45 lb/ac/day (5 $gm/m^2/day$) in December 1965 in the control ponds. When the productivities of the control were compared to the productivities of the test and nutrient addition ponds, the data indicated that (1) harvested hyacinths removed sufficient nutrients to retard future productivity if the harvested nutrients were not returned to the water and (2) the harvested hyacinths removed nutrients in sufficient quantities to retard future production when additional nutrient supplements were added to the ponds. Perhaps more importantly this work provided the best estimate of monthly productivity for an entire year of water hyacinth growth in Florida. It was noted that when harvesting is employed, there was a net productivity during every month of the year. This is definitely contrary to water hyacinth productivity in a natural unharvested environment in central Florida as noted by Penfound and Earle [23] and is essential if water hyacinths are to provide a consistent nutrient removal alternative. The most optimistic growth estimates are by Wolverton and McDonald [36] and Dinges [12] who have estimated that from 500-1600 lb/ac/day (56-180 gm/m^2/day) dry weight can be produced by water hyacinths grown in warm sewage effluents.

WATER QUALITY PARAMETERS AFFECTING GROWTH

Chadwick and Obeid [8] working at the University of Khartoum in England grew hyacinths in tap water and Long Ashton solution at varying pH values of 3.0, 4.5, 5.6, 6.9, and 8.2 in a laboratory environment. The growth solutions were continuously aerated and adjusted for volume and pH control. Their results indicate that the yield of water hyacinths in either solution would be approximately optimized at pH 7. Chadwick and Obeid [8] further demonstrated that an increase in plant mass corresponds to an increase in the number of plants which does not reduce the mass of the original plants. The greatest yield of plant material was obtained in the nutrient fortified solutions. These results agree with Parra[22] as noted by Chadwick and Obeid[8]. Parra[22] measured water hyacinth growth in tap water that was adjusted to varying pH values. Parra's results also indicated that optimum plant growth occured approximately at pH 7. In natural stands of water hyacinths the pH of the underlying water is most likely to e acidic due to the absence of sunlight and resulting reduction in photosynthetic activity. Penfound and Earle[23] report that on the average, pond waters in the Mississippi River Delta have a pH of approximately 7.2 but where the water is near or around large mats of water hyacinths the pH ranges from 6.2 to 6.8. Recently Wolverton and McDonald[33] reported a pH decrease in raw sewage effluent passing through a 0.70 acre lagoon from 7.5 to 6.6 for a 3 month period. Haller and Sutton[17] investigated the effect of pH on water hyacinth growth and found that the optimum plant growth ranged from pH 4.6 to 7.3. The highest yield was experienced at pH 4.6, which is unusual, but their technique of removing the plants during the growth period could have retarded growth as noted by Yount and Crossman[38].

The natural cycle of water hyacinth appears to lower pH by reducing photosynthetic activity by light reduction.

Penfound and Earle[23] reported that complete coverage of water hyacinth infestation changes the physical factors of the underlying water. Typically the surface water temperature is more uniform, the pH is lower, the CO_2 content is higher and the dissolved oxygen content is lower. They summarize their findings in relation to the amount of decaying matter present in the stands. Dissolved oxygen concentration under heavy mats with decay was less than 0.1 ppm, under heavy mats without decay was less than 0.5 ppm, in open mats 1.5 ppm and in adjacent open pools the dissolved oxygen content was 4.0 ppm. In performing a five month colonization study of water hyacinths, Penfound and Earle[23] found that the plant doubling rate was fifty days in a pool constructed in the midst of a large mat of water hyacinth where the dissolved oxygen concentration was 0.8 ppm. The same technique produced a plant doubling rate of approximately 12 days when undertaken in an open pool where the dissolved oxygen concentration ranged from 3.5 to 4.8 ppm.

Bock[4] reported that plants were found in the Old River in Jamaica growing in 60% full sunlight. She was unable to find any plants growing in less than 60% full sunlight and subsequent work by Bock[4] indicated that water hyacinths will not survive in less than 50% full sunlight. Sixteen plants nurtured in a greenhouse died when the light intensity was 30-40% of full sunlight in a six month period. Heavily overgrown swamps could possibly eliminate the threat of additional infestations of water hyacinths used in water treatment if the water flowing from the plants was channeled through the swamp.

Penfound and Earle[23] conducted batch experiments to determine the effect of salinity of water hyacinths. They placed five plants each in various dilutions of salt water from the Mississippi Sound. The results of their study are given in Table 7 and indicate that the plants could only survive in 0.06% salinity or 600 mg/l for periods greater than 28 days. Many sewage effluents have total dissolved solids concentrations of 1000 mg/l or greater. The toxic effect of excess salinity in sewage may limit the use of hyacinths in areas where they would be otherwise acceptable for nutrient removal. However, the plants have been successfully grown in many waters consisting of or receiving sewage effluents, Furman[7], Wolverton and McDonald McDonald[33 34 36] and Ornes and Sutton[21]. Bock[4] observed water hyacinths in river waters with 30% salinity. This river did eventually empty into the sea but the mats of hyacinths were seemingly static.

TABLE 7.

MAXIMUM AND MINIMUM LAND REQUIREMENTS
FOR A WATER HYACINTH NUTRIENT REMOVAL SYSTEM
BASED ON NET PLANT PRODUCTIVITY AREA

Nutrient	Maximum ac/mg/1/mgd	Medium ac/mg/1/mgd	Minimum ac/mg/1/mgd
Nitrogen	8.78	1.53	0.79
Phosphorus	36.26	6.30	3.26

Based on a plant dry weight content of 2.1% N and 0.5% P.

Water hyacinths being highly productive floating aquatic plants with a relatively large surface area have high rates of transpiration. Timmer and Weldon[30] studied the evaporation and transpiration rates of water hyacinths in Fort Lauderdale, Florida from late April to early September from man-made pools containing canal water. The results of their work, when statistically analyzed, implied a high degree of correlation between the cumulative solar radiation and cumulative water lost due to evapotranspiration for the study. The evapotranspiration rate averaged 3.96 inches/week as opposed to an evaporation rate of 1.08 inches/week, or 3.7 times the evaporation rate. This is significant for communities in water-short areas which depend on surface water supplies that are fed by drainage basins infested with water hyacinths. Melbourne, Florida is a good example of local community experiencing this problem. The loss of water due to plant infestations would exceed 6 acre/feet of infested area. In the Upper St. Johns River Basin the infested acreage would range possibly into the thousands of acres if not more. Because these plants are not harvested or controlled they do not improve water quality and remove large amounts of water during critical supply periods. In addition, water hyacinths adversely affect water quality by limiting access and by depositing large amounts of organic matter can deposit up to five or seven inches/year of sediment, Penfound and Earle[23].

There is evidence that water hyacinths create a biological system that will effectively remove nitrogen in the NO_3^{-2} or NH_4^+ form. Dunnigan, Phelan and Shamsuddin[13] measured NO_3-N and NH_4^+ changes in the field and laboratory from waters that were supporting water hyacinth growth. Their results showed no significant differences in the nitrogen form on uptake in the lbaoratory but water hyacinths in the field were found to have more effect on the decrease of NH_4^+ than the NO_3^{-2}-N concentration in the water media. Significant losses of both forms of nitrogen were recorded in either instance which indicate these plants could be used to effectively treat the effluent out of a trickling filter or an activated sludge plant. Sheffield[28] operating an 8 liter/day pilot plant in Orlando, Florida that consisted of a water hyacinth pond which had a detention time of ten days, an NH_3 air stripping unit and a coagulation process was able to remove 99% of either NO_3^{-2} or NH_3-N. He recommended operating the pond in an anaerobic condition to promote denitrification but noted that a phosphorus release from decomposing sediments occurred that decreased the initial phosphorus removal rate. An important point is that in periods of low productivity the hyacinth ponds could be operated in an anaerobic condition to promote nitrogen removal through denitrification instead of plant uptake.

APPLICATIONS

Little data has been published on pilot plant studies involving water nyacinths for nutrient removal. Work by Furman[16] at the University of Florida utilized water hyacinths for removing nutrients from sewage effluent from a secondary treatment plant. Through plant ponds, employing varying depths and detention times, nitrogen removals of 80% were obtained. Cornwell, et.al.[9] published data from the University of Florida study relating the percent nitrogen or phosphorus removal to the surface loading rate and expressed this relation by the following equations:

(2) $$PR = 0.238\ (SA) - 8.77$$

(3) $$NR = 0.327\ (SA) + 6.95$$

PR = Percent Phosphorus Removal
SA = Thousands of Sq. Ft./mgd
NR = Percent Nitrogen Removal

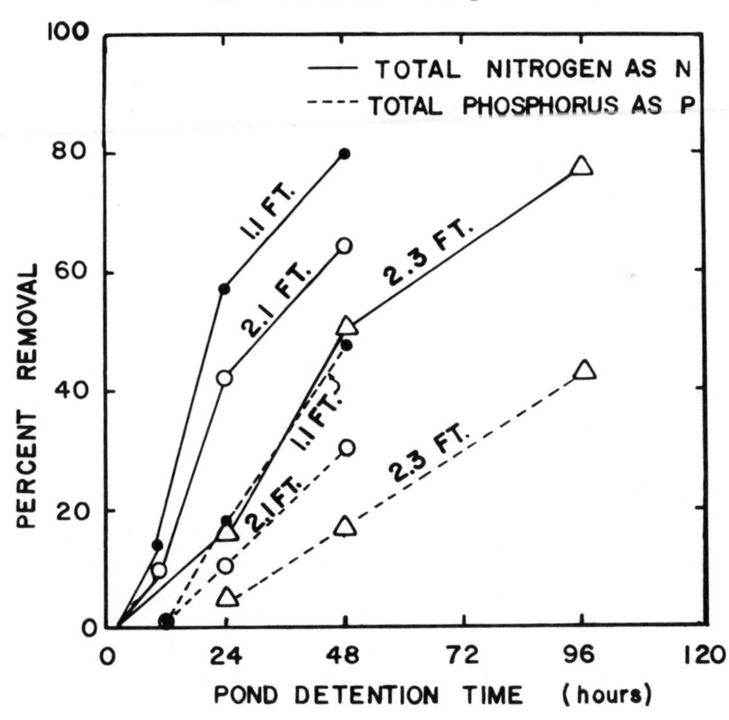

FIGURE 5. Total Nitrogen and Total Phosphorus Removal in Small Test Ponds as a Function Corresponding to Detention Time and Depth

REFERENCE: Cornwell, D.A., Zoltek, J., Jr., Patrinely, C.D., Furman, T. des, and Kim, J.I. "Nutrient Removal By Water Hyacinths". Copyright Vol. 49, and reprinted with permission.

The percent nutrient removal that corresponded to pond detention time and depth is shown in Figure 5. The average total nitrogen and phosphorus in the effluent was 13.68 and 3.44 mg/l respectively, with the majority of each of the ortho P or NO_3^--N forms. Cornwell, et.al.[9] suggested that surface area and depth were related to nutrient uptake in a direct manner, such that if pond depth increased a corresponding increase would have to occur in pond surface area to maintain the same nutrient removal.

The molar ratio of N/P removal in the University of Florida study was 8.7/1 giving a weight removal ratio of 3.93/1. This is somewhat less than the average N/p weight ratios for water hyacinths reported by Parra and Hortenstine[22] of 5.2/1 and indicates that the nitrogen removal could be attributed to plant growth. One of the unique points about this work is that it represents data for all seasons and it is the only study that has systematically tried to develop design data for the use of water hyacinths for nutrient removal.

Wolverton and McDonald[33 34] have published results from investigations covering a two year period. In the first study conducted at Gulport, Mississippi water hyacinths were used to upgrade sewage lagoons that cover 0.70 acres and received an inflow of 0.115 gpd from an aerated primary sewage lagoon. The detention time in the lagoon ranged from 14 to 21 days depending on the transpiration rate of the plants. Data was collected from hyacinths from July through September. The suspended solids content of the effluent from the water hyacinth lagoon averaged 9 mg/l or 83% less than the suspended solids concentration from the control lagoon. This was attributed to the decline of algal production in the hyacinth lagoon. Decreases in ammonia nitrogen were also noted but the change in total nitrogen in either lagoon was not determined. In Bay St. Louis, Mississippi water hyacinths were allowed to fill 2.5 ha of a 17.5 ha lagoon. Previous to the induction of water hyacinths in the lagoon several problems had resulted from excessive algal blooms and decomposition causing high suspended solids in the effluent and offensive odors that offended nearby residents. The suspended solids concentration dropped from a high of 148 mg/l in May to a low of 8 mg/l in October. The average suspended solids concentration in the effluent was 15 mg/l once the 2.5 ha coverage was complete which resulted in approximately a 90% reduction in suspended solids concentration. These applications of water hyacinths were for secondary waste treatment. At Bay St. Louis, the BOD reduction was not greatly inproved by the addition of water hyacinths to the lagoon but the influent was relatively weak (39.25 mg/l) for a domestic waste. However, in the water hyacinth filled lagoon at Gulfport the effluent BOD was reduced 84% relative to the hyacinth free lagoon.

Wolverton and McDonald[34] have reported a 90% suspended

solids reduction and a 95% BOD reduction for a 4.94 acre primary lagoon covered with water hyacinths. The lagoon received an incoming flow of 0.126 mgd of raw sewage containing 110 mg/l BOD and 97 mg/l suspended solids which was retained for 54 days. The BOD and suspended solids loading were 23.31 lb BOD/ac/day and 20.4 lb SS/ac/day. This is the only data known to the authors which would provide a means to size a hyacinth treatment facility for BOD or suspended solids removal. Total nitrogen resuction was not reported. Total phosphorus was reduced 56% from 3.7 to 1.6 mg/l. The hyacinths were not harvested at a rate to optimize nutrient removal. The average phosphorus uptake was 2.1 lb/ac/day which is approximately twice that attributed to phosphorus uptake from hyacinth growth, Yount and Crossman[38] and Parra and Hortenstein[22]. This indicates that the water hyacinths are not directly responsible for all water quality changes but do provide a major environmental change in the ecosystem which allows additional nutrient uptake.

The cost effectiveness of any system that employs water hyacinths for nutrient removal will depend in part on the net productivity and nutrient content of the plant. Ornes and Sutton[21] and Younts and Crossman[38] have shown that the net productivity can vary from 45 to 259 lb/ac/day (5-29 gm/m^2/day) for water hyacinths that are periodically harvested. While the most involved investigator utilizing water hyacinths in wawaste treatment systems, Wolverton[36], has estimated that water hyacinths can achieve a dry weight yield of 500 lbs/ac/day (56 gm/m^2/day). The nitrogen and phosphorus content of these plants can range from 1.61% to 3.8% N and from 0.31% to 1.0% P, Parra and Hortenstein[22]. Boyd[6], Nasa[35], and Boyd and Blackburn[5].

The necessary land area can be calculated for any desired nutrient removal if the net plant productivity and nutrient content are assumed and balanced against the nutrients available in the secondary effluent by equation 4.

$$A = 8.34Q_i N_1/(1-C_1)K_1C_2 \quad (4)$$

A = acres/mgd
Q_i = incoming flow, mgd
N_1 = nutrient concentration of influent, mg/l
C_1 = percent H_2O of plant composition
K_1 = net plant productivity, lb/ac/day
C_2 = percent nutrient concentration of dry weight

From equation 4 and the ranges previously given for net productivity, %N and %P, the maximum and minimum land requirements were determined that would provide 1 mg/l of nutrient removal from an incoming flow of 1 mgd and a nutrient concentration of 1 mg/l. The areas in Table 7 are given in ac/mg/l/mgd so that the land area could be quickly approximated once the desired nutrient concentration of the waste stream was selected.

For example, if 1 mgd of wastewater contained 20 mg/l total nitrogen and an effluent of 10 mg/l total nitrogen was desired, the land requirement would range from 7.9 to 87.8 acres. These removal rates are based solely on hyacinth growth and harvesting. Enhanced nutrient removal may occur as a result of ecosystem interactions not directly attributable to water hyacinth growth but brought about by their influence.

The cost effectiveness of any treatment system employing water hyacinths for nutrient removal will depend on the productivity and nutrient content of the plant. Previous work has demonstrated that the rate of plant growth is affected by pH, dissolved oxygen, temperature, space and the available nutrient concentrations of the plant containing waters. Feasibility studies by Robinson et. al.[26] have suggested that in Florida water hyacinth systems are more economical for nitrogen removal than conventional AWT techniques.

No projects known to the authors have been executed that attempted to maximize water hyacinth productivity in nutrient removal systems. Studies have been completed that employed these plants for nutrient removal but none of these varied the previously mentioned factors to increase productivity which would increase nutrient removal if proper harvesting rates are utilized. It is highly likely that a water hyacinth removal system would require pH control and lagoon aeration to maximize net productivity. Lack of control of these and other factors are probabily the reasons for the high variability of existing data for water hyacinth productivity. In periods of low productivity, lagoons could be operated in an anaerobic state to promote nitrogen removal by denitrification as opposed to harvesting.

Basic information needs to be gathered for a water hyacinth process from a system's viewpoint on the denitrification, limiting nutrient concentrations, BOD reduction processes, seasonal productivity, and seasonal harvesting rates. Research and demonstration projects are required to gather this information and transfer it into basic engineering data that will allow a water treatment system to be properly evaluated and designed for waste treatment.

COST

When evaluating the cost of water hyacinth systems, cost for pond construction, pond management, plant harvesting, plant processing and plant handling must be considered. While there are several methods of processing and handling water hyacinths, only three are being considered in this evaluation. These are composting with sludge, drying for carrle feed, and pressing with subsequent disposal to a landfill. Other possibilities include methane gas production, pyrolysis, incineration,

incineration, livestock bedding, poultry feed, ethanol production and paper production. Methane production and paper production have been investigated to some degree[1,34] the other areas are still in need of extensive research. The total cost for the water hyacinth systems considered in this paper is given by equation 5. A simplified flow diagram of the conceptual water hyacinth systems considered in this paper is given by equation 5. A simplified flow diagram of the conceptual water hyacinth systems is shown in Figure 6.

$$(5) \quad P = A_c + A_{om} + B_c + B_{om} + a(C_c + C_{om}) + b(F_c + F_{om}) + c(L_c + L_{om})$$

where:
- P = total system present worth in dollars
- A_c = capital cost of treatment system prior to hyacinth ponds in dollars
- A_{om} = O&M present worth of the treatment system prior to hyacinth ponds in dollars
- B_c = capital cost of hyacinth ponds and harvesting equipment
- B_{om} = O&M present worth of the hyacinth pond and harvesting system
- C_c = capital cost of composting system
- C_{om} = O&M present worth of the composting system
- F_c = capital cost of feed production system
- F_{om} = O&M present worth of the feed production system
- L_c = capital cost of the pressing and landfill system
- L_{om} = O&M present worth of landfill system

If $a = 1$, $b = 0$ $c = 0$
If $b = 1$, $a = 0$ $c = 0$
If $c = 1$, $a = 0$ $b = 0$

This gives the total cost of a hyacinth system. Description and development of each of these variables in equation 5 is required before the equation becomes useful. The following system baseline is taken from existing literature and is used to make this cost estimate.

Baseline

General Economic Considerations

* Costs are EPA STP Construction Index of 260
* Twenty year analysis period
* Cost analysis will be based on present worth
* Interest rate of 6-3/8 percent

General Systems Considerations

* Land required for supporting facilities and buffer zone are calculated as 15 percent of the pond area

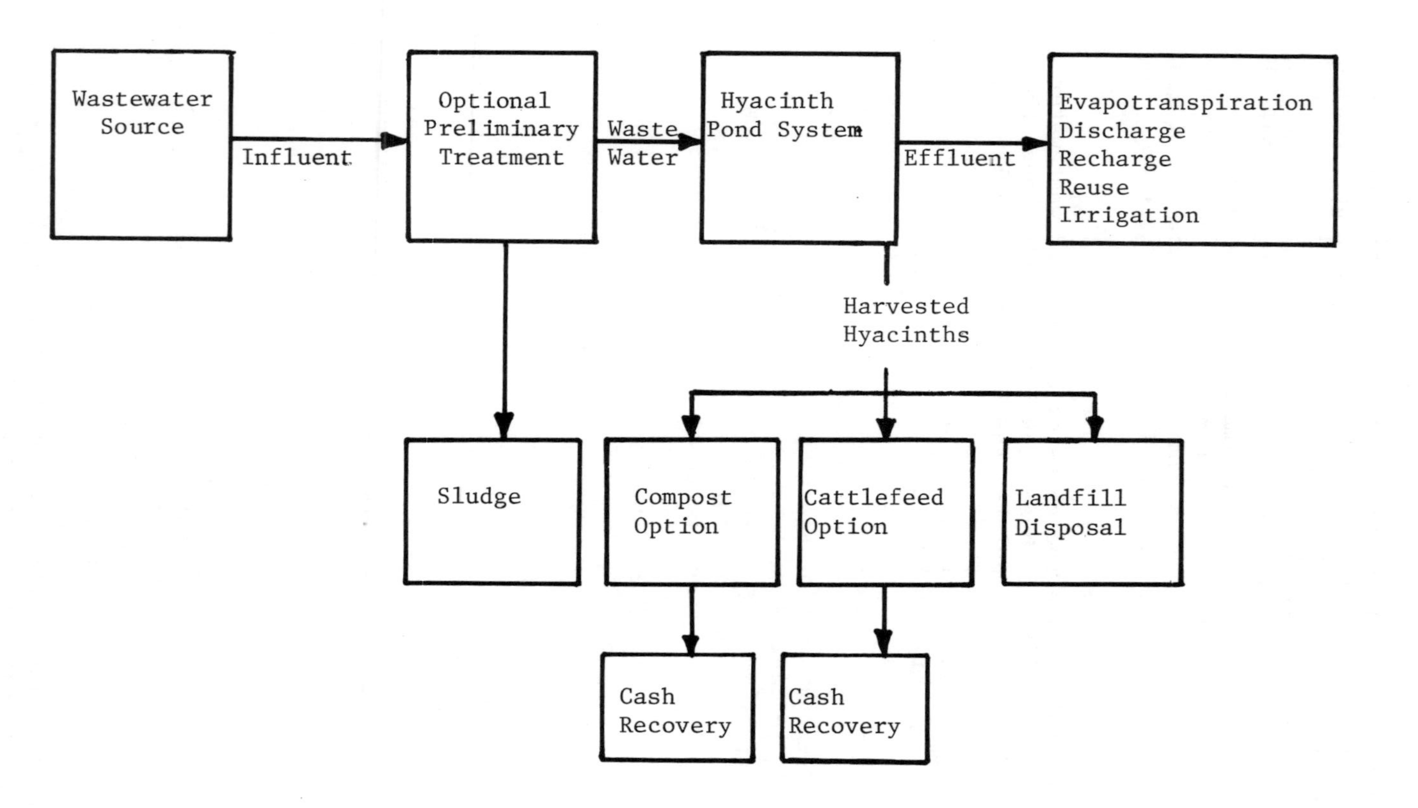

FIGURE 6. Flow Diagram of Waste Treatment System Utilizing Water Hyacinths

*Hyacinths are harvested at 3 wet tons/acre-day
*Pond water depth is four feet
*Excavation costs are $\$1.25/yd^3$ (Richardson[25])
*Clay lining is $\$5,000/acre^2$
*Harvesting is done by a 3 hp barscreen at a maximum rate of 12 tons/hour[24]
*Annual maintenance costs for the barscreen are set at 2 percent of capital cost[24]

A_c AND A_{om}

The A_c and A_{om} are the costs of any desired wastewater treatment system prior to hyacinths. If there is none, these variables are zero, as would be the case in a water hyacinth lagoon receiving raw sewage. The capital equipment and O&M costs for secondary treatment may be obtained from equations 6, 7, and 8.

$$(6)\ A_c = 1.77Q^{.696}$$

$$(7)\ A_{om} = (38-27.3 \log_{10}Q)(3650\ Q \times pwf) \quad \text{If } Q \leq 10 \text{ mgd}$$

$$(8)\ A_{om} = (15.8-5.1 \log_{10}Q)(0.365\ Q \times pwf) \quad \text{If } Q \leq 10 \text{ mgd}$$

where:
A_c = capital costs of any wastewater treatment system preceding a water hyacinth system in millions of dollars
A_{om} = operation and maintenance costs of any waste treatment system preceding a water hyacinth system in millions of dollars
Q = average daily flow in mgd
pwf = present worth factor, = 11.39 for 6 3/8 percent interest over 20 years

B_c is determined from equation 9 and provides the cost of the land, construction, lining and harvesting.

$$(9)\ B_c = (1-sf)(1.15)T_a l_c + 1613T_a\ d\ E + \frac{1.10\ H_w T_a h}{R} + T_a\ s$$

where:
T_a = total acreage needed for hyacinth ponds
sf = present worth salvage factor
l_c = cost of land \$/acre
d = depth of cut in feet
E = excavation cost \$/cubic yard
H_w = hyacinth production in wet tons/day
R = harvesting rate per harvester in tons/day
h = cost (4) of each harvester
s = cost for sealing ponds \$/acre

The factor 1.15 provides for the buffer zone requirements. The factor 1613 converts acre-ft to cubic yards. In the third term, 1.10 expands the number of harvesters required by 10 percent for backup, replacement, etc. To determine the total acreage T_a, equations 10 through 13 are required.

$$T_{ab} = \frac{8.34Q\ (B_i - B_e)}{K_b} \quad (10)$$

$$T_{as} = \frac{8.34Q\ (S_i - S_e)}{K_s} \quad (11)$$

$$T_{an} = \frac{8.34Q\ (N_i - N_e)}{0.01\ N_h\ H_d} \quad (12)$$

$$T_{ap} = \frac{8.34Q\ (P_i - P_e)}{0.01\ P_h\ H_d} \quad (13)$$

where:

T_{ab} = acres required for desired BOD removal
T_{as} = acres required for desired suspended solids removal
T_{an} = acres required for desired nitrogen removal
T_{ap} = acres required for desired phosphorus removal
B_i = BOD_5 of influents to hyacinths pond, mg/l
S_i = suspended solids influents to hyacinths pond, mg/l
N_i = total nitrogen influents to hyacinths pond, mg/l
P_i = phosphorus influents to hyacinth pond, mg/l
B_e = desired BOD_5 of effluent of mg/l
S_e = desired suspended solids of effluent in mg/l
N_e = desired total nitrogen effluent in mg/l
P_e = desired phosphorus effluent in mg/l
K_b = BOD_5 removal rate in lb/acre/day
K_s = suspended solids removal rate in lb/acre/day
N_h = percent nitrogen content of hyacinth as dry weight
P_h = percent phosphorus content of hyacinth as dry weight
H_d = annual average hyacinth dry weight productivity in lb/acre/day
8.34 = conversion from mg/l to lbs/mgd

The following conditions can now be set for T_a.

(14) $$T_a = \max \{T_{ab}, T_{as}, T_{an}, T_{ap}\}$$

This based design for the most demanding criteria which would theoretically better all other effluent criteria.

B_{om}

B_{om} is the sum of the energy costs for harvesting, the labor costs for harvesting, the maintenance cost and the labor cost for management which is given by equation 15.

(15) $$B_{om} = \frac{(6528\ H_w T_a h_p)}{R}\ S_{kwh}\ pwf + (MH_b\ HR + S_m)\ pwf$$

where:
- h_p = horsepower of harvester
- S_{kwh} = Cost \$/kwh
- MH_b = total annual man hours
- HR = hourly wage \$/hr
- S_m = maintenance cost/\$/yr (usually 2-5 percent of equipment cost)
- 6528 = conversion from kwh/yr to hp-day/day

C_c and C_{om}

The composting operation herein is a rapid 5 day method similar to Metro-Waste System.[19] Composting is proceeded by a belt press sludge dewatering system into which both hyacinths and sludge are fed. This dewatering system has capital cost of \$75,000/ton dry solids and operating costs of \$10/dry ton.[11] This includes costs for pumping, polymers, labor and power. This cost of the composting operation was approximated from preliminary cost data on the Metro-Waste composting system presented in Table 10.[19] Recovery costs from compost sales range from \$20/ton to \$35/ton.[19]

The composting process utilizes ordinary sludge as well as pressed water hyacinths. Any waste treatment system that involves composting would not require a digestor. Digestor costs must be eliminated from normal secondary treatment costs before these costs can be used to accurately estimate costs of any waste treatment process utilizing composting. Composting costs were developed from data supplied by Resource Recovery, Inc., Houston, Texas.[19] Energy and labor costs were estimated to be \$0.40/kwh and \$10/hour, respectively.

The capital costs of the water hyacinth system following secondary treatment and utilizing composting was determined by summing the costs of the belt press and the composting equipment less the cost savings on digestors. The construction costs were approximated at \$60,000/dry tons-day,[24] and are presented in Table 9. The capital cost for composting are represented by equation 16.

TABLE 9.

APPROXIMATE METRO-WASTE COMPOSTING COSTS
FOR A HYACINTH-SLUDGE OPERATION

Dry Tons Produced Daily (input)		Construction Cost Million $	Labor Costs $/yr	Energy and Maintenance $/yr
Sludge	Hyacinths			
14.4	38.3	2.9	58,400	110,000
17	52.8	4.0	58,400	151,000
17	0	1.2	29,200	48,000

20-Year Present Worth O&M Million $	Tons/Day Recovery at 40% Water[10]	Annual Million Dollars Recovered		20-Yr Present Worth of Recovery	
		$20/ton	$35/ton	$20/ton	$35/ton
1.92	87.0	0.64	1.12	7.29	12.76
2.39	114	0.83	1.46	9.45	16.63
0.88	28	0.2	0.36	2.39	4.10

$$(16) \quad C_c = [\frac{(T_a \times H_d) + (Q \times Z)}{2000}](S_{bp} + S_{co}) - Y\,A_c$$

where: Z = lbs/mgd of dry sludge
S_{bp} = capital costs of belt presses/dry tons of solid/day
S_{co} = capital costs of composting equipment/dry tons of solids/day
Y = digestor cost fraction of secondary treatment cost

The annual operating cost of the composting system is the sum of the energy costs, labor costs, and maintenance costs less the recovery costs from the sale of the product. As shown in Table 9, the energy, maintenance, and labor costs for the composting operation are $90/dry ton, while the belt presses cost $10/dry ton to operate. The composting operating and maintenance costs were determined from equation 17.

$$(17) \quad C_{om} = 365\ \text{pwf}\{[\frac{(T_a x H_d) + (QxZ)}{2000}][O_{bp}+O_c]-[\frac{(1-L_s)(T_a x H_d x G)}{2000}]\}$$

where: O_{bp} = $O&M cost/dry ton for belt press
O_c = $O&M cost/dry ton for composting
L_s = fraction of dry solids lost in composting
G = = selling price $/ton of 40% moisture compost

F_c and F_{om}

Harvested water hyacinths reduced from 95% to 15% moisture are potentially marketable as cattle feed at $25/ton as an ingredient of a total cattle feed pellet.[14] Bagnall, et.al[1] had previously explored the use of hyacinths as a cattle feed. They found that its worth is enhanced by reduction of moisture content and by mixing with other materials so that it comprises no more than 30 percent of the total diet. To reduce the moisture content of the hyacinths, Bagnall[2] has designed a screw press which can process a maximum of 28 tons per hour. The construction cost of this press is $15,000.[2] The press operates at about 2.5 horsepower hours/wet ton or 1.85 kwh/wet ton,[1] which amounts to $0.07/wet ton based on an energy cost of $0.04 kwh.

After pressing, the hyacinths must be dried before they can be utilized for cattle feed. Bagnall[2] and Wolverton[35] have both investigated the use of solar driers. Baird and Bagnall[3] found that the average drying time (to 15 percent moisture content) was 128 hours in a direct insolation, downflow drier. This was an open system using pressed hyacinths loaded at 2.5 lb/ft^2 and a down draft of 5 cfm/ft^2. The area required was 2,122 ft^2/ton of hyacinths which had been reduced to an 80% moisture content by pressing. The cost for the solar drier system was estimated at $1.00 per sq. ft.[3] The capital costs of the cattle feed system utilizing water hyacinths is given in equation 18.

$$(18) \quad F_c = \frac{1.10\ T_a x H_w x S_{sp}}{J} + TT_a H_d t_d \left[S_{sd} + \frac{2.3x10^{-5}(1-sf)1_c}{0.20\ M}\right]$$

where:
- S_{sp} = cost $/screw press
- J = wet tons/day handled by the screw press
- S_{sd} = cost $/sq ft for solar driers
- M = solar drier loading rate in lbs/sq ft
- 0.20 = conversion of dry plant weight to plant weight at 80 percent moisture
- t_d = total drying time in days

The operation and maintenance cost are given by equation 19 and consist of the labor, energy and maintenance costs less the market return on the sale of the compost.

$$(19) \quad F_{om} = (MH_f \times HR + S_{mf})pwf + \frac{6528\ H_w H_{pp} \times T_a \times S_{kwh} \times pwf}{J} - \frac{H_d \times T_a \times 365 G_f \times pwf}{1700}$$

where: MH_f = manhours/year needed for processing
S_{mf} = annual maintenance cost for presses and driers ($/yr)
h_{pp} = horsepower of presses
G_f = sale price of dried hyacinths ($/ton)
1700 = conversion of lbs dry weight to tons containing 15% moisture

L_c and L_{om}

The final processing option deals with the hyacinth harvest as a solid waste problem. Costs include pressing to reduce weight, hauling and landfill costs. Hauling costs include both truck capital and O&M costs. The hauling costs of a 12-ton truck for a ten-mile haul are $1.70/ton as estimated by Liptak.[18] Landfill charges are estimated to be $2.59/ton. The capital cost of a 12-ton truck with a five-year life expectancy is approximately $35,000. The capital cost and O&M cost were determined from equations 20 and 21, respectively.

$$L_c = \frac{1.10\ H_w\ S_{sp}\ T_a}{J} + (1 + SPF_5 + SPF_{10} + SPF_{15})\left(\frac{1.1\ S_{tr}\ H_w\ T_a}{0.80\ W\ \ V}\right) \tag{20}$$

where: SPF_5 = single payment factor at five years
SPF_{10} = single payment factor at ten years
SPF_{15} = single payment factor at fifteen years
S_{tp} = $ cost/truck
W = weight capacity of the truck in tons/trip
V = trips taken/day for each truck

$$L_{om} = (MH_{pl})pwf + \frac{6528 H_w\ T_a\ h_{pp}\ S_{kwh}\ pwf}{J} + \frac{S_h H_d T_a I\ 365\ pwf}{400} + \frac{S_{lf} H_d T_a\ 365\ pwf}{400} \tag{21}$$

where: MH_{pl} = manhours/year for pressing and loading
S_h = hauling cost in $/ton-mile
I = round trip miles/trip
S_{lf} = $/ton landfill cost

HYPOTHETICAL APPLICATION

A municipality is involved in evaluating various wastewater treatment and effluent disposal options. The existing secondary facility is essentially nonfunctional and must be replaced. The predicted flow for a twenty year horizon is 10 mgd and phasing is not being considered at this time. The regulatory agency has given an effluent discharge allocation of 3 mg/1-BOD, 3 mg/1-SS, 0.6 mg/1-TN, and 0.3 mg/1-TP. They have also stated that secondary treatment would be adequate for land spreading at the rate of 125 acres/mgd. Investigations into landspreading operation reveals a suitable site 5 miles away. The site is prime development land which sells for $10,000/acre. The land cost, considering a salvage factor of 0.291 (6-3/8%, 20 years) is $5,849,000 with a ten percent safety factor. The consultants seeing the difficulties that might arise with the land spreading option decided to investigate hyacinths as a means of waste treatment. From pilot study and marketing investigations the design parameters were determined and are given in Table 10.

TABLE 10.

PARAMETERS ESTABLISHED, FOR EXAMPLE, 10 mgd HYACINTH TREATMENT FACILITY

Parameter	Description	Value
l_c	land cost	$10,000/acre
pwf	present worth factor	11.39
sf	salvage factor	0.29
E	excavation costs	$1.25/cubic yard
H_w	rate of wet weight hyacinth production	2.5 tons/acre-day
H_d	rate of dry weight hyacinth production	250 lbs/acre-day
R	harvesting rate/harvester	120 tons/day (10 tons/day)
r	harvesting rate/harvester	12 tons/hour
h	cost of each harvester (barscreen)	$21,000
s	cost for pond lining	$5,000/acre
d	depth of excavation cut	1.5 feet
B_i	BOD influent to ponds	200 mg/1 raw, 20 mg/1 secondary
B_e	BOD effluent	3 mg/1
S_i	SS influent to ponds	200 mg/1 raw, 20 mg/1 secondary

TABLE 10 (Continued)

Parameter	Description	Value
S_e	SS effluent	3 mg/l
N_i	TN effluent to ponds	25 mg/l
N_e	TN effluent	0.6 mg/l
P_i	TP influent to ponds	10 mg/l
P_e	TP effluent	0.3 mg/l
k_b	BOD removal rate	23.3 lb/acre-day
k_s	SS removal rate	20.4 lb/acre-day
N_h	percent N of hyacinth dry weight	2.1 percent
t_d	solar drying time	5.3 days
P_h	percent phosphorus of hyacinth dry weight	0.5 percent
h_p	horsepower of harvester	3
S_{kwh}	cost of electricity	$0.04/kwh
MH_b	man hours/year for harvesting & pond	1/2 of harvesting time + 2080
HR	hourly wage	$10/hour
S_m	annual maintenance of ponds and harvester	2% of harvester costs
Z	raw sludge production	1600 lbs/mgd
S_{bp}	capital cost of belt presses	$75,000/dry ton solids/day
S_{co}	capital cost of composter	$60,000/dry ton solids/day
Y	fraction of secondary cost for sludge digestion	0.32
O_{bp}	O&M cost, belt presses	$10/dry ton
O_c	O&M cost, composting	$9/dry ton
L_s	fraction of solids lost in composting	0.20
G_c	price of compost	$25/ton
S_{sp}	cost of screw press	$15,000
J	handling capacity of screw press	280 wet tons/day
S_{sd}	cost of solar drier platform	$1.00/ton
M	solar drier loading rate	2.5 lbs/sq. ft.
MH_f	manhours for processing	1/2 pressing time + 4160 hours
S_{mf}	annual maintenance for presses	4 percent of capital
h_{pp}	horsepower of presses	70 hp
SPF_5	single payment factor, 5 years	0.734

TABLE 10 (Continued)

Parameter	Description	Value
SPF_{10}	single payment factor, 10 years	0.539
SPF_{15}	single payment factor, 15 years	0.396
MH_{pl}	manhours for pressing	pressing time
S_{tr}	cost of truck	$35,000
W	weight capacity of truck	12 tons/trip
V	trips taken daily	2/day
S_h	hauling costs	$0.17/ton-mile
I	round trip distance	20 miles
Q	flow	10 mgd
G_f	price of cattle feed	$25/ton
S_{lf}	landfill cost	$2.59/ton

After developing these parameters, equations 6 through 21 were used to estimate each process cost and the total system cost. The hyacinth ponds were preceded with a conventional secondary system (because of land limitations). Therefore, B_i and S_i of 20 mg/l were used in the calculations. The lagoon area required for the water hyacinth system was determined from equation 14. If a secondary treatment facility had not preceded the water hyacinth system, a removal rate of 23.3 lbs BOD_5/acre-day[36] and 20.4 lb SS/acre-day[36] would have been used to determine land requirements. The nitrogen and phosphorus removal rates were based on a plant productivity of 250 lbs dry weight/acre-day which contained 2.1% N and 0.5% P. Table 11 gives the values derived from the cost equation.

From the following cost analysis, the most cost effective water hyacinth system required marketing the harvested plants as cattle feed. The present worth of the water hyacinth systems were $\$52.9 \times 10^6$ for land spreading, $\$36.6 \times 10^6$ for compost and $\$21.2 \times 10^6$ for cattle feed as an alternate means of disposal. The cattle feed alternative hypothetically was optimum because it required a low capital investment and returned a profit, negative present worth. The system that involved composting was projected to be less costly than the final system which involved landfilling which had no marketable outlet for the harvested hyacinths. The 20 year present worth of a 10 mgd advanced waste treatment plant was estimated to be $\$26.3 \times 10^6$ from the cost data published by the State of Florida Department of Environmental Regulation in the "Guide to State Approval of 201 Facility Plans." AWT facilities typically are projected to provide a finished water

TABLE 11.
CALCULATION OF COSTS FOR THE EXAMPLE 10 mgd HYACINTH FACILITY

Parameter	Description	Equation Number	Value
A_c	conventional secondary capital cost	6	$\$8.79 \times 10^6$
A_{om}	O&M PW conventional secondary	7	$\$4.45 \times 10^6$
T_{ab}	land required for BOD removal	10	60.8 acres
T_{as}	land required for SS removal	11	69.5 acres
T_{an}	land required for TN removal	12	388 acres
T_{ap}	land required for TP removal	13	647 acres
T_a	land required for hyacinth growth	14	647 acres
B_c	capital cost of hyacinth ponds and harvestors	9	$\$10.80 \times 10^6$
B_{om}	O&M PW cost of ponds and harvesting	15	$\$3.11 \times 10^6$
C_c	capital cost of composting	16	$\$9.18 \times 10^6$
C_{om}	O&M PW cost of composting	17	$\$0.30 \times 10^6$
F_c	capital cost for feed production	18	$\$0.33 \times 10^6$
F_{om}	O&M PW cost for feed production	19	$\$6.36 \times 10^6$
L_c	capital cost loading	20	$\$8.75 \times 10^6$
L_{om}	O&M PW cost landfill	21	$\$17.04 \times 10^6$
P at a=1 b,c=0	total system, marketing plants as compost	5	$\$36.6 \times 10^6$
P at b=1 a,c=0	total system, marketing plants as cattle feed	5	$\$21.2 \times 10^6$
P at c=1 a,b=0	total system, disposal of plants in landfill	5	$\$52.9 \times 10^6$

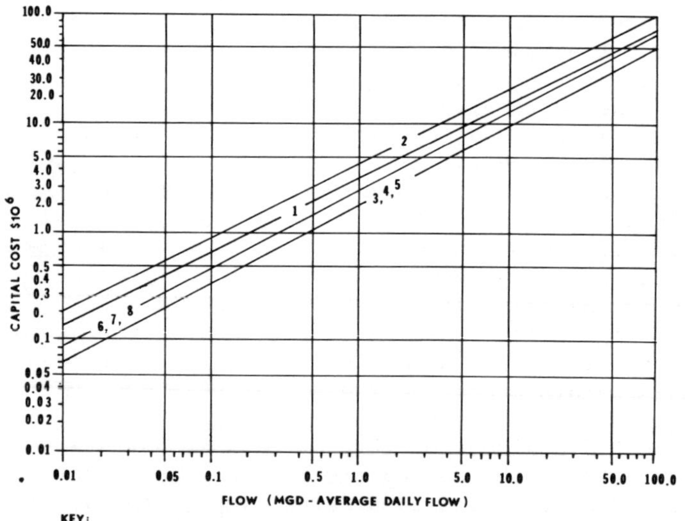

FIGURE 7. Estimated capital costs for hyacinth treatment systems as compared to conventional wastewater treatment systems.

of 5/5/3/1 and would be stressed to comply with the natural background quality of most receiving waters. Consequently, landspreading is currently strongly recommended for a waste treatment system. The cost of a typical land spreading system (2"/wk) is estimated to be $\$26 \times 10^6$,[11] and does not compare favorably with the water hyacinth system involving the harvested plant as cattle feed.

The estimates for the capital and O&M costs of water hyacinth systems involving landfilling, marketing as compost or cattle feed are given in figures 7 and 8 as a function of flow, mgd. These cost curves were generated from equations 5 through 21 but were modified to represent total capital outlay. The capital costs of a hyacinth system are controlled by the required acreage/mgd which is controlled by productivity and nutrient content of the plant. The O&M costs however are somewhat independent of plant productivity and are estimated as a function of the number of wet tons of plant material harvested per mgd. The plant nutrient content would affect this but the plant productivity is not as critical. The reasoning being the same mass of nutrients has to be removed no matter what the productivity. It is not to be inferred that the O&M costs of the three plant options will not be different but that the harvest rate in wet tons/mgd is a more reasonable basis for O&M estimation.

Several alternatives are considered in figures 7 and 8. Basically the costs of the three different hyacinth systems are compared against land spreading following secondary treatment and AWT. Capital costs for the hyacinth systems were determined at 20 and 70 acres/mgd. These rates correspond to plant productivities of 500 and 142 lbs dry weight/ac-day. Based on a plant dry weight content of 2.1% N and 0.5% P, the nitrogen standard of 0.6 mg/l N would be met but the standard of 0.3 mg/l P would not be met solely by plant productivity. Approximately 83.4 lbs P/mgd would be discharged to the hyacinth pond which would result in 50.5 lbs P incorporated by the plant biomass.

Based on plant productivity alone, the phosphorus concentration in the effluent would be 3.94 mg/l. However, existing literature[34 9] has documented that more phosphorus removal occurs in water hyacinth lagoond than can be accounted for by plant productivity. In Mississippi[34] approximately twice as much P was removed and at the University of Florida[9] 32% additional P was removed than could be accounted for by plant productivity. The data presented in figures 7 and 8 are based on the assumption that an additional 36% P removal would result from on site precipitation and absorption in the pond. This figure could well be increased by the addition of colloidal limestone that would provide nucleating sites for the phosphate precipitate and is an excellent research project.

The different plant productivities used in figures 7 and 8 provide and estimated cost range of the hypothetical hyacinth

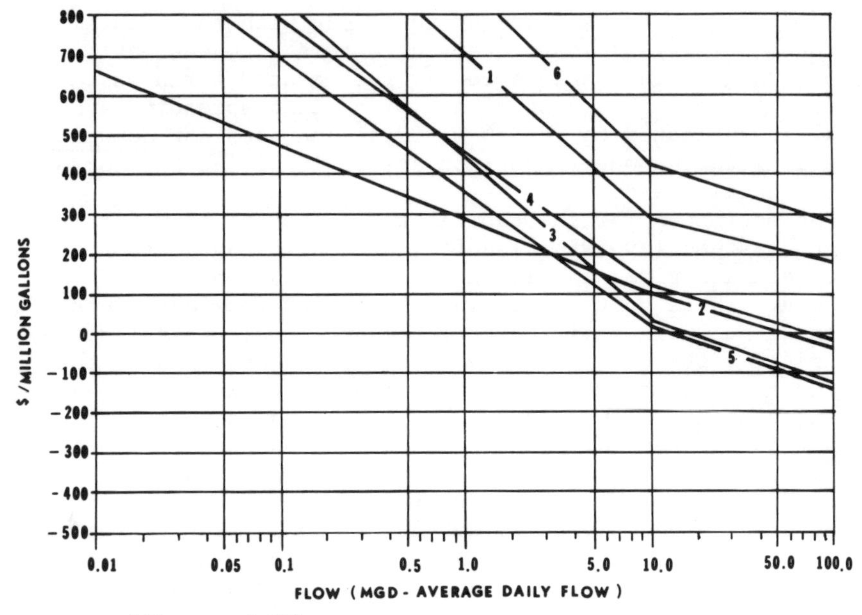

FIGURE 8 Estimated O&M costs for hyacinth treatment systems as compared to conventional wastewater treatment systems.

systems. Although there are some slight differences in the capital equipemnt costs of the three different plant disposal systems, they do have a significant bearing on cost projections. Therefore, only one curve was developed to estimate the capital equipment costs for the three different water hyacinth systems. The O&M costs are minimized for all flows up to 3 mgd when a system requiring land spreading following secondary treatment is specified. The optimum system for flows greater than 3 mgd requires water hyacinths and marketing of the the harvested plants as cattle feed.

The total cost of any of the systems can be estimated by determining the present worth of the capital cost and summing it to the O&M costs. For the systems described, the high costs of capital equipment outweighs the O&M costs and suggests the hyacinth systems would be more economical at 20 or 70 acres/mgd with either marketable option. These curves can be used to estimate the approximate cost of any hyacinth system.

It is recommended that individuals first use these curves to determine if water hyacinth systems are feasible in a given locale. If they are, then it is strongly recommended that each individual use equations 5 through 21 to develop costs that would more accurately estimate his environment.

These equations do provide a systematic method of estimating costs of a potential waste treatment system that could represent considerable savings. The cost curves shown in figures 7 and 8 do provide a cost estimate but should not be used in place of equations 5 through 21 when estimating costs of water hyacinth waste treatment systems.

It is not the intent of this paper to imply that water hyacinth systems are always more cost effective then either AWT or landspreading but rather to demonstrate that large-scale demonstrations and research projects are needed and that these systems are potentially very competitive in tropical or subtropical regions.

SUMMARY

The market for hyacinths must be carefully examined. In this analysis, it is evident that if solar driers become practical, then using hyacinths as cattle feed is attractive. However, the product needs further study before it can be considered a commodity. Composting may be the best method in many cases, for it handles sludge also. Sludge disposal is a major part of wastewater treatment costs, so the benefits of its conversion into a marketable product can not be ignored. Sludge compost is presently being sold in Florida.[19] Other possible products from hyacinths are methane gas, which has been researched by both Bagnall[1] and Wolverton[37], and horse bedding, a use to which dried hyacinths is quite well suited. This study shows that disposal of hyacinths into landfills is not cost effective except for possibly small flows (less than 1 mgd). Therefore, development of a marketable outlet for the harvested plants is mandatory for the system to be cost effective unless than land requirements can be greatly reduced.

Another consideration is that of harvesting. As noted, early investigations on harvesting techniques involved the use of draglines which cost from $125,000 to $250,000 each and require about $35/hour to operate. Draglines harvest only about 8 tons/hour. For example, the barscreen method for a 10 mgd facility would have a present worth of $1,002,625. The present worth of draglines would be $38,694,000. Selection of a proper harvesting technique is obviously a critical consideration in designing a hyacinth system, and an area for future research.

At this time, little is understood about the dynamics involved in nutrient uptake within a hyacinth system. In this analysis it has been assumed that nutrients are removed only by the plants. However, hyacinths are associated with a complex food web which is involved in the movement and storage of nutrients. Insect larvae, for example, upon maturation leave the system, thereby accounting for some loss. Mosquito fish and other fish are harvested on a regular basis by birds and other predators. There is undoubtedly considerable phosphorus loss to the sediments. Nitrogen may be lost through denitrification under anaerobic conditions. The haycinth pond is an ecosystem well adapted to high nutrient loads. Learning how to maximize its ability to extract these nutrients is the most crucial factor in determining the future viability of hyacinth systems in wastewater management. Such practice, for example, as genetic selection of the more efficient plants might enhance productivity. Intermittent anaerobiosis may enhance nitrogen removal. Aeration, covering or heating the influent, may also help increase productivity. The use of larval forms such as tadpoles may increase nutrient removal. Precipitation with aluminum salts might well be integrated into the system. Certainly an investigation of an overall mass balance model which maximizes nutrient removal would help determine the direction of such management practices and provide needed information for successful management of water hyacinth systems.

The cost presented here are intended to aid one in evaluating hyacinths for their particular situation. The four items which must be carefully addressed during such a study are land costs, annual productivity rates, harvesting methods and cost effectiveness of development of a market for the harvested plants. Once these are established, it is possible to determine the feasibility of a nutrient removal system utilizing water hyacinths. Whether or not the hyacinth system is selected depends upon the actual costs developed at that time. Because of the low operations cost, high recovery potential and low energy demands, hyacinths represent a potential wastewater treatment technology that with proper investigation and management might well serve tropical or subtropical areas around the world.

REFERENCES

1. Bagnall, L.O., T. deS. Furman, J.F. Hentges, Jr., W. J. Nolan, and R. L. Shirley. (1974) Feed and Fiber from Effluent-Grown Water Hyacinth, In. Wastewater Use in the Production of Food & Fiber--Processing. EPA-660/2-74-041. Washington, D.C.
2. Bagnall, L.O. (1977) University of Florida, Gainesville. Agricultural Engineering. Personal Communication.
3. Baird, C.D. and L.O. Bagnall. (1975) Solar Crop Drying in the Sunshine State. American Society of Agricultural Engineers. 1975 Annual Meeting. Davis, CA.
4. Bock, J.H. (1966) An Ecological Study of Eichhornia Crassipes with Special Emphasis on its Reproductive Biology. University of California, Berkeley, CA.

5. Boyd, C.E. and R.D. Blackburn. (July 1970) Seasonal Changes in the Proximate Composition of Some Common Aquatic Weeds. HCJ, Vol. 8, No. 2.
6. Boyd, C.E. (1970) Vascular Aquatic Plants for Nutrient Removal. Economic Botany, Vol. 24.
7. Boyd, C.E. and R.D. Blackburn. (1970) Seasonal Changes in the Proximate Composition of Some Common Aquatic Weeds. Hyacinth Control Journal 8:42-44.
8. Chadwick, M.H. and M. Obeid. (1966) Eichhornia crassipes and Pistia stratiotes. The Journal of Ecology. 54:563-575.
9. Cornwell, D.A., J. Zoltek, Jr., C.D. Patrinely, T. deS. Furman, and J.I. Kim. (Jan. 1977) Nutrient Removal by Water Hyacinths, JWPCF, Vol. 49, No. 1.
10. Corps of Engineers, Savannah District. (1975) Wastewater Treatment Unit Processes, Design and Cost Estimating Data. Atlanta.
11. Dawkins and Associates, Inc., Orlando, FL. Consulting and Environmental Engineers. Recent Cost Investigations.
12. Dinges, W.R. (1976) Who Says Sewage Treatment Plants Have to Be Ugly. Water and Wastes Engineering. Vol. 13, No. 4.
13. Dunigan, E.P., R.A. Phelan and A.M. Shamsuddin. (1975) Use of Water Hyacinth to Remove Nitrogen and Phosphorus from Eutrophic Waters. Hyacinth Control Journal. 13:59-62.
14. Florida Cattle Feed Producers. Personal Communication.
15. Florida Department of Environmental Regulation. (1977) Guide to State Approval of 201 Facility Plans. Tallahassee.
16. Furman, T. deS. (1971-77) University of Florida. Personal Communication.
17. Haller, W.T. and D.L. Sutton. (1973) Effect of pH and High Phosphorus Concentrations on Growth of Water Hyacinth. Hyacinth Control Journal. 11:59-61.
18. Liptak, B.G. (1974) Environmental Engineers Handbook. Chilton Books. Radnor, PA.
19. Myrick, H.N. (1978) Resource Conversion Systems, Inc. Houston, TX. Personal Communication.
20. Quote from Orange County Sanitary Landfill. (1977) Orlando, FL.
21. Ornes, W.H. and D.L. Sutton. (1975) Removal of Phosphorus from Static Sewage Effluent by Water Hyacinth. Hyacinth Control Journal. 13:56-58.
22. Parra, J.V. and C.C. Hortenstein. (1974) Plant Nutritional Content of Some Florida Water Hyacinths and Response by Pearl Millet to Incorporation of Water Hyacinth in Three Soil Types. Hyacinth Control Journal. 12:85-80.
23. Penfound, W.T. and T.T. Earle. (1948) The Biology of the Water Hyacinth. Ecol. Mono. 18:447-478.
24. Rice, R. (1978) President, Dyneco, Inc., Lauderhill, FL. Personal Communication.
25. Richardon Engineering Services, Inc. (1977) Process Plant Construction Estimating Standards. Volume 4. Solana Beach, CA.

26. Robinson, A.C., H.J. Gorman, M. Hillman, W.T. Lawhon, D.L. Maase and T.A. McClure. (Jan. 1976) An Analysis of the Market Potential of Water Hyacinth Based Systems for Municipal Wastewater Treatment. Report No. 3CL-OA-TFR-76-5, Battelle, Columbus, OH.
27. Scarsbrook, E. and D.E. Davis. (1971) Effect of Sewage Effluent on Growth of Five Vascular Aquatic Species. Hyacinth Control Journal. 13:59-61.
28. Sheffield, C.W. (June 1976) Water Hyacinth for Nutrient Removal. Hyacinth Control Journal, Vol. 6.
29. Sutton, D.L. and R.D. Blackburn. (1971) Uptake of Copper by Water Hyacinth. Hyacinth Control Journal. 13:59-61.
30. Timmer, C.E. and L.W. Weldon. (1967) Evapotranspiration and Pollution of Water by Water Hyacinth. Hyacinth Control Journal, Vol. 6.
31. United States Environmental Protection Agency. (1974). Process Design Manual for Sludge Treatment and Disposal. EPA 625/1-74-006. Washington, D.C.
32. Wahlquist, H. (1972) Production of Water Hyacinths and Resulting Water Quality in Earthen Ponds. Hyacinth Control Journal, Vol. 10.
33. Wolverton, B.C. and R.C. McDonald. (Oct. 1975) Water Hyacinths for Upgrading Sewage Lagoons to Meet Advanced Wastewater Treatment Standards: Part I. TMX-72729, NASA, NSTL, Bay St. Louis, MO.
34. Wolverton, B.C. and R.C. McDonald. (Oct. 1976) Water Hyacinths for Upgrading Sewage Lagoons to Meet Advanced Wastewater Treatment Standards: Part II. TMX-72730, NASA, NSTL, Bay St. Louis, MO.
35. Wolverton, B.C. and R.C. McDonald. (Feb. 1978) Personal Communication to E. A. Stewart.
36. Wolverton, B.C. and R.C. McDonald. (1976) Water Hyacinths for the Removal of Phenols from Polluted Waters. Aquatic Botany, 2.
37. Wolverton, B.C. (1977) Research Scientist National Space Technology Laboratory. Bay St. Louis, MO. Personal Communications.
38. Yount, J.L. and R.A. Crossman, Jr. (1970) Eutrophication Control by Plant Harvesting. JWPCF, Vol. 42.

9. CYPRESS SWAMPS FOR NUTRIENT REMOVAL AND WASTEWATER RECYCLING*

Katherine Carter Ewel. School of Forest Resources and Conservation, and Center for Wetlands, University of Florida, Gainesville, Florida

H. T. Odum. Department of Environmental Engineering Sciences, and Center for Wetlands, University of Florida, Gainesville, Florida

INTRODUCTION

Cypress domes appear to be ecologically and economically suitable for treatment of secondary sewage effluent. Tree growth rates are substantially increased and understory vegetation is not significantly disturbed. Neither nitrogen nor phosphorus concentrations in the groundwater surrounding the experimental wells increase. Nor would there be greater export of encephalitis virus from the domes than would be expected from an undisturbed swamp. Because the ratio of solar energy to dollar costs that it uses is high, disposal of wastewater into wetlands is a good energy conservation practive and is economically cost effective.

Many communities throughout the country are facing increases in utility costs as new water pollution laws come into effect. Installation of advanced wastewater treatment systems is prohibitively expensive for smaller communities

*Summary of work done by a research group through the University of Florida Center for Wetlands supported by grants from The Rockefeller Foundation and the Division of Applied Science and Research Applications of the National Science Foundation.

in particular. A project undertaken by the Center for Wetlands at the University of Florida has been investigating for nearly five years the consequences of use of cypress wetlands for sewage recycling. Secondarily treated sewage has been discharged at the rate of 2.5 cm/wk into two cypress domes, which are small swamps commonly found in pine flatwoods and plantations. The main study area (Figure 1) is located near Gainesville, Florida, where three of the four experimental cypress domes are located next to a trailer park. Two of these were accidentally burned in a fire early in the project; one has been receiving sewage and another groundwater since March, 1974. A third, unburned dome has been receiving sewage since March, 1975. A fourth dome located in the university's Austin Cary Forest is surrounded by a natural stand of slash pine and longleaf pine. No experimental treatments have been applied to this dome.

CYPRESS ECOSYSTEMS IN FLORIDA

Cypress trees are commonly found where water levels fluctuate: in floodplain forests, around the fringes of lakes, in domes, in extensively meandering, shallow, slowly moving streams called strands, and in flat, poorly drained areas where stunted scrub cypress grow. The greatest extent and density of cypress trees in north-central Florida occur in domes and strands, which are often underlain by clay[1]. Peat and limestone of varying depths are found beneath these ecosystems in south Florida, however, where optimum growth rates for cypress occur in strands at a hydroperiod of 286 to 296 days[2]. In this area, wetland habitats occur in areas with hydroperiods of at least 223 days. Peat depth and consequently vegetation composition and growth rates may vary if fire burns through an area with any frequency, however. Early growth rates may be faster where peat is deepest[2], but trees rooted in these areas may be more easily killed by fires[3].

Gross primary productivity rates are often very low in cypress domes but have been found to be equal to or higher than Puerto Rican rain forest rates in floodplain forests and in one of the sewage-enriched domes[4]. Nutrient levels and resulting growth rates are dependent in part on the area of land that drains into a swamp. This area is lowest for scrub cypress, greater but highly variable for domes, still greater for strands, and most for floodplain forests[4]. Hydroperiod also affects productivity. Growth is maximum when the hydroperiod is long enough to reduce competition but not so long as to eliminate other vegetation entirely. Accordingly, Figure 2 shows that cypress that grows in association with pine, and cypress in pure stands represent these extremes. Cypress in these ecosystems grow more slowly than cypress in association with hardwoods, which are capable

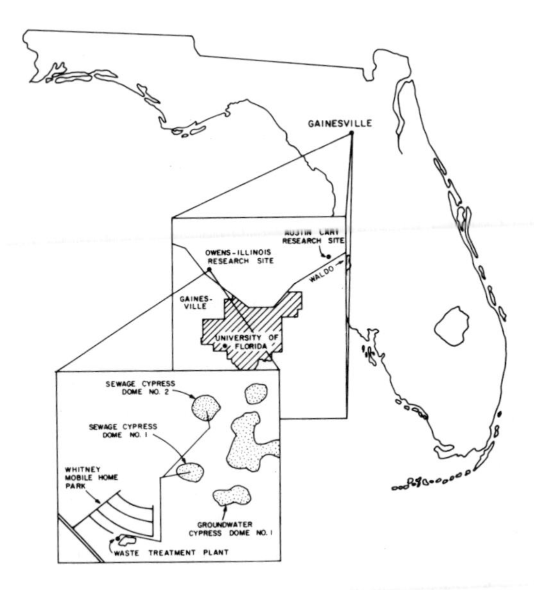

FIGURE 1. Map of Sewage Recycling Research Sites

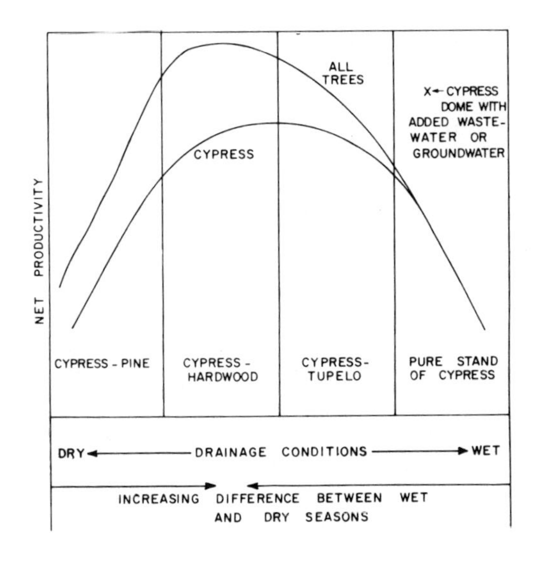

FIGURE 2. Relationship Between Net Productivity and Drainage in Several Wetland Ecosystems[5]

of withstanding a moderate hydroperiod[5].

EFFECT OF SEWAGE ON VEGETATION

Cypress dome vegetation is usually distinct from vegetation in the surrounding pine flatwoods. Species composition appears to be affected primarily by fluctuating water levels. Ferns are common in the shallower edges of the dome, small floating plants such as duckweed may grow in the deeper interior pool, and shrubs and herbs may be rooted in the organic matter that accumulates around the bases of trees and on cypress knees.

Application of secondarily treated sewage to cypress domes increases biomass of understory vegetation (Figure 3), primarily because of the proliferation of small, floating plants (Lemna perpusila, Spirodela oligorhiza, and Azolla carolinensis)[6]. Bald cypress, pond cypress, and black tupelo seedlings planted in the cypress domes showed varying degrees of success. Seedlings of pond cypress, which is the variety of cypress found in domes, grow at approximately the same rate in both Sewage Dome 1 and the Groundwater Dome (Figure 4). Because these domes had been previously burned, light penetration to the water surface is greater than in the other sites. Growth rates of seedlings planted at Austin Cary are significantly lower. The mortality rate of seedlings planted in the experimental domes was considerably higher in the sewage domes than in either of the control domes[7]. Increases in diameter of mature trees, on the other hand, have been greater in the experimental domes than in the control dome (Table I). Although the effects of the fire have confounded the results, the concentrations of nutrients in trees appear to increase initially but to return eventually to normal[8].

Similarly, increased productivity rates in sewage-enriched cypress domes are apparently due more to increases in leaf biomass than to increases in weight-specific photosynthetic rates[4].

TABLE I. AVERAGE[a] ANNUAL INCREASE[b] IN DIAMETER OF TAXODIUM DISTICHUM VAR NUTANS (6/26/76-6/15/77)[8]

Sewage Dome 1[c]	Sewage Dome 2[d]	Groundwater Dome 1[e]	Austin Cary Control Dome[f]
0.33 ± 0.03	0.36 ± 0.03	0.36 ± 0.03	0.08 ± 0.01

[a]Values are the mean and plus or minus the standard error of the mean

[b]Values listed are in centimeters

[c]Values are based on 94 trees

[d]Values are based on 97 trees

[e]Values are based on 98 trees

[f]Values are based on 195 trees

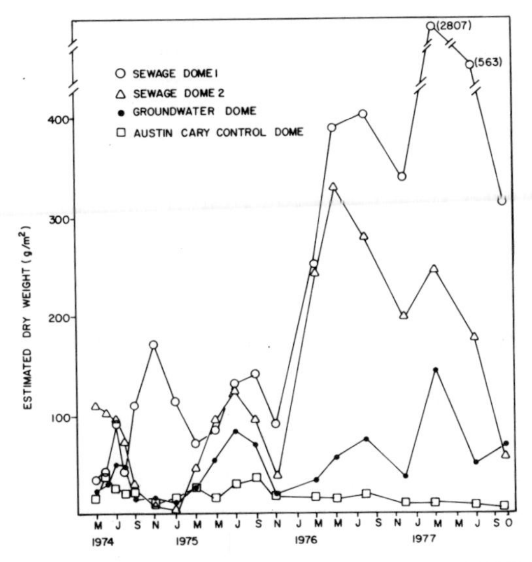

FIGURE 3. Changes in Biomass of Understory Vegetation[6]

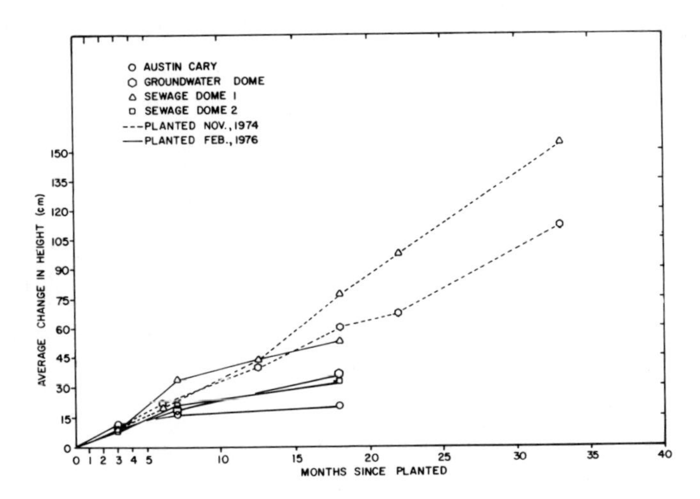

FIGURE 4. Growth Rates of Pond Cypress Seedlings Planted in Four Experimental Cypress Domes[7]

NITROGEN AND PHOSPHORUS RELATIONSHIPS

Phosphorus budgets calculated for the unburned sewage dome and for the control dome are shown in Figure 5[9]. In the sewage dome, at least 72% of the incoming phosphorus is estimated to percolate from the surface water downward; in a short-term experiment with radioactive phosphorus, 9% of this amount was found to be retained in the layer of organic material lining the basin. Increased uptake of phosphorus by faster-growing trees only accounts for 2% of the incoming phosphorus in this study, leaving approximately 25% of the incoming phosphorus unaccounted for. However, in an analysis of a nearby sewage-enriched cypress strand, high concentrations of phosphorus were found in the roots of cypress[10]. In this swamp, it was estimated that as much as 43% of the phosphorus taken up by the ecosystem may have been stored in the root tissue.

Neither nitrogen nor phosphorus appears to be moving laterally from underneath the domes. The presence of chloride ions in shallow wells surrounding the domes indicates that the water entering the dome from the sewage treatment plant is infiltrating the shallow water table[11]. However, the sediments, sands, and clays underlying the dome are retaining many of the other elements, particularly nitrogen and phosphorus (Figure 6).

Dierberg and Brezonik[11] conclude that nitrogen rather than phosphorus is more limiting in the sewage-enriched cypress domes. They present four observations to support their conclusions: 1.) low N:P ratio relative to the ratio for the control dome; 2.) variability of nitrogen levels over time relative to phosphorus; 3.) the nitrogen in the dome existing in inorganic form; and 4.) correlation of BOD with total nitrogen.

ANIMAL POPULATIONS IN CYPRESS DOMES

Figure 7 illustrates the effect of sewage disposal into cypress swamps on leopard frogs: the sewage domes act as a sink, whereas the groundwater dome, representing an experimental control, is an exporter of amphibian biomass[12]. Fish populations are greatly reduced, but are seldom an important component in cypress domes because of fluctuating water levels and low aquatic productivity. Diversity of bird species was found to be greater in a sewage dome than a control dome[12]. Moreover, the number of bird sightings in the sewage dome was 150% greater than the number of sightings in the control dome. Part of this increase is due to higher numbers of migrating birds.

An extensive sampling effort showed that insect biomass was greater in the sewage domes than in the control domes,

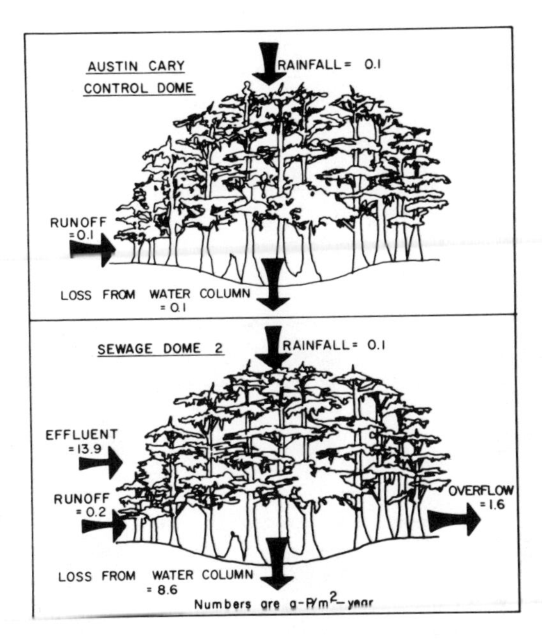

FIGURE 5. Major Phosphorus Flows in a Natural Cypress Dome and in a Sewage-Enriched Cypress Dome[9]

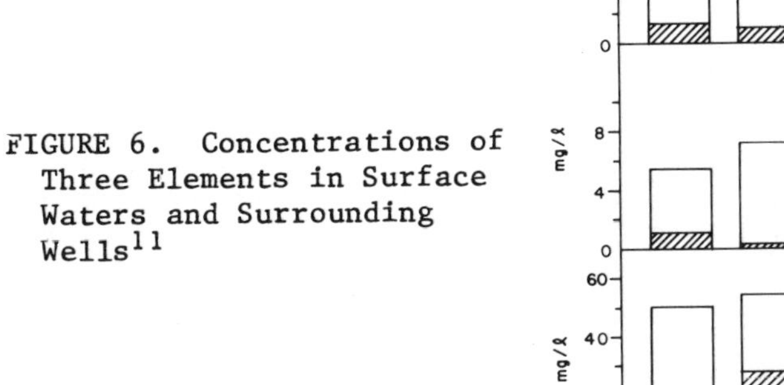

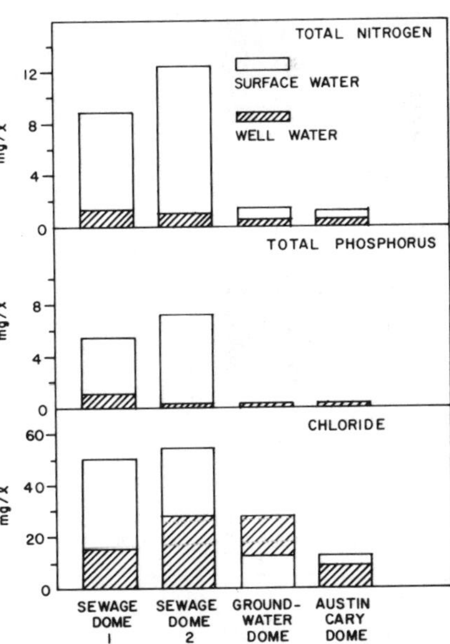

FIGURE 6. Concentrations of Three Elements in Surface Waters and Surrounding Wells[11]

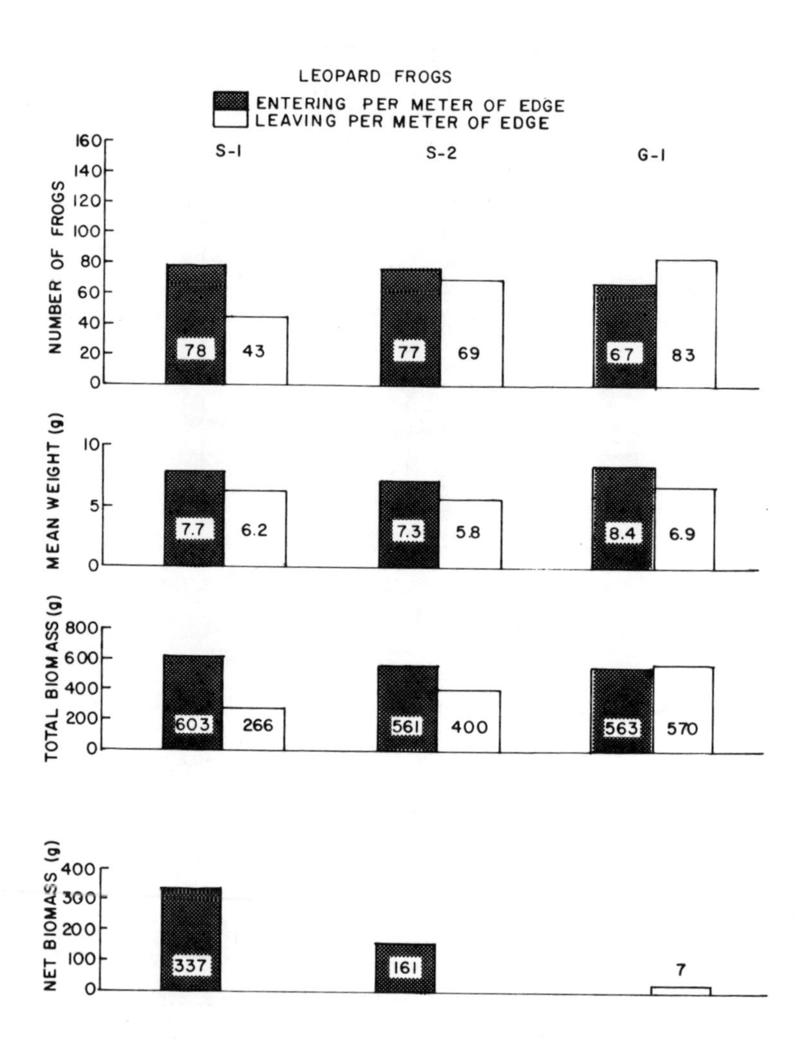

FIGURE 7. Numbers and Biomass of Frogs Captured at the Edges of Three Experimental Cypress Domes[12]

although all domes show a high level of diversity with little similarity between them[13]. The diversity was as high as values measured in the Puerto Rican rain forest using the same technique. No significant differences in mosquito populations, particularly of economically or medically important species, were found between the domes. Nor did the sewage domes produce significantly higher levels of eastern or western equine encephalitis virus than other domes. Most of the virus activity is during the summer. However, 96% of the birds sighted in both the sewage dome and the natural dome were residents and winter visitors. The summer visitors comprised only 3% of the total in both cases. Export of encephalitis virus from the domes would therefore not be significantly greater than export from an undisturbed dome. Fecal coliform levels in groundwater wells in the experimental domes and in the standing water at a nearby sewage-enriched strand were consistently low during a 1.5-year sampling program[15].

FEASIBILITY

A preliminary study of the cost of three different methods of advanced wastewater treatment for the city of Waldo, Florida, resulted in the following estimates[16]:

Spray Irrigation	$0.63/1000 gal
Advanced Waste Treatment	1.07/1000 gal
Wetland Recycling	0.42/1000 gal

These values do not include the cost of secondary treatment. Waldo is a small city, and now produces 20 to 30 thousand gallons of sewage per day. The estimates were based on an anticipated flow of 120,000 gallons per day by 1990. The wetland recycling scheme proposed impoundment of part of a larger nearby cypress wetland.

A more general analysis[17] indicated that the extensive force main network and fencing costs needed for disposal from large sewage treatment plants into many small cypress domes would make disposal into domes more expensive than spray irrigation. They also found, however, that the cost of discharge into a single large cypress strand was competitive with spray irrigation for these larger disposal systems. Research into the structure and function of cypress strands is being continued along with the ongoing cypress dome work in order to analyze the ability of these ecosystems to provide an important service to society that might be compatible with their normal ecological roles.

THE ROLE OF SWAMPS IN SAVING WATER

A hydrologic budget of an unburned cypress dome showed 30% less water lost as evapotranspiration than from an open water surface (Figure 8)[18]. A similar study done in a cypress strand came to the same conclusion[19]. A different pattern is seen in agricultural systems: increases in productivity of a plant population increase water loss over evaporation from bare soil [20, 21]. Some cypress swamps, however, have low leaf area indices and leaf biomass[4, 22]. Moreover, cypress trees drop their leaves in the dry season, stopping transpiration, while still shading the water and diminishing wind strength, thereby keeping evaporation rates low.

By transpiring less, cypress trees help maintain their own characteristic wetland habitat. The draining of cypress ponds in Florida is causing a loss of water that would otherwise be available for aquifer recharge or for economic use. A government memorandum circulated in Florida a decade ago advocated cutting swamp trees to save water. If implemented, this would clearly have hurt the economy of Florida.

ENERGY ANALYSIS

The role of cypress dome recycling can be measured with methods used in energy analysis. In Figure 9, the energy embodied in the work of the swamp is shown as an input from the left (I). Renewable energies that are available directly and indirectly from the sun help to treat the sewage without economic cost. To connect and process the wastewaters in a manufactured sewage treatment plant requires purchased goods and services. These contain embodied energy from the main economy in proportion to the money spent. The ratio of economic cost to free environmental services, both expressed as coal equivalents, is 11.5:3, a factor of 3.8. This is only slightly larger than the average ratio of fuel use to solar-based energy (in coal equivalents) in the U.S., which is about 2.5.

From an environmental-protection point of view, this is an acceptable ratio, because it does not increase the amount of economic activity relative to environmental subsidies above the average value for the U.S. From an economic point of view, the contribution of free, renewable energy flow in the environment is substantial, providing matching energy and thereby attracting economic activity to help make it competitive with systems requiring greater economic input.

Technological tertiary treatment, for instance, has large dollar costs without much contribution from environmental processes. The combined system of man and nature is competitive ecologically and economically when the two aspects are reinforcing and cooperating. The economy benefits and

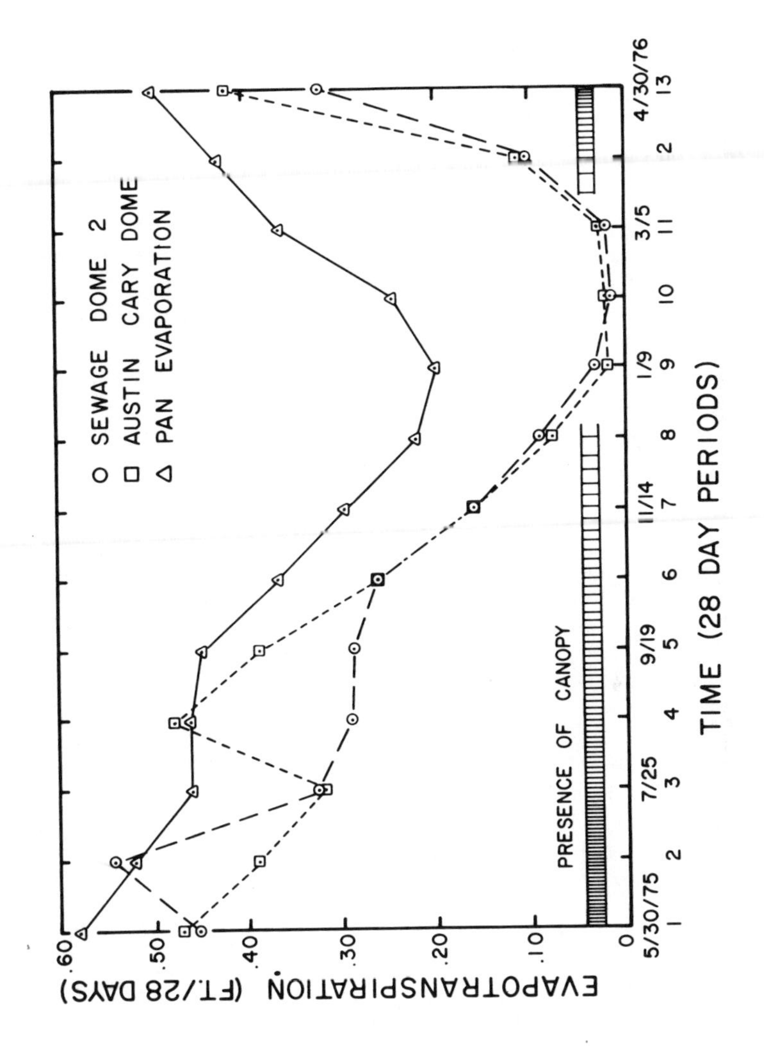

FIGURE 8. Seasonal Evapotranspiration Rates from Two Cypress Domes, and Corresponding Pan Evaporation Rates[18]

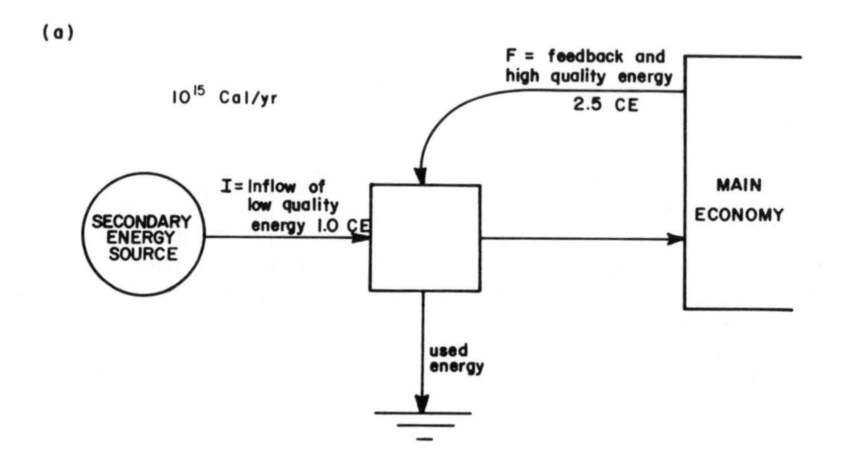

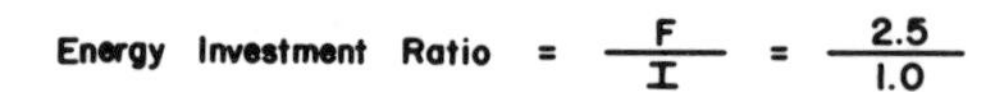

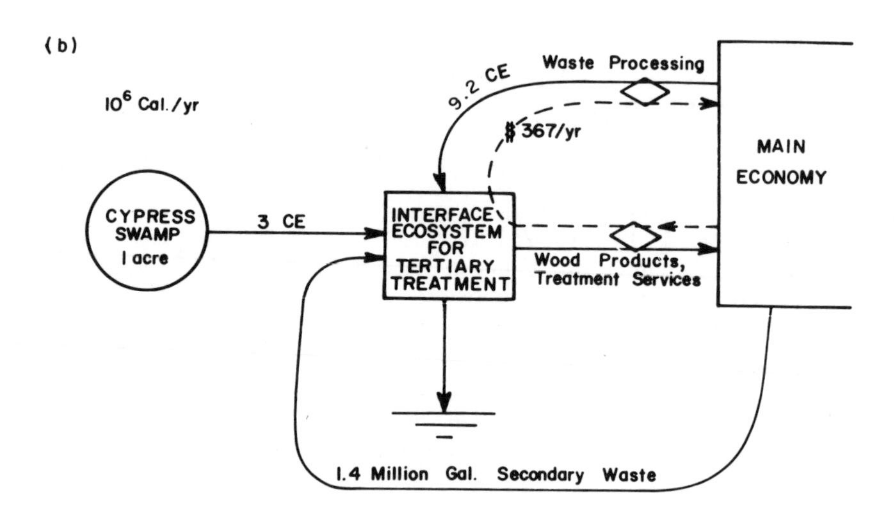

FIGURE 9A. CALCULATION OF UNITED STATES INVESTMENT RATIO USING COAL EQUIVALENTS. HIGH QUALITY ENERGY PUMPS AN INFLOW OF LOW QUALITY ENERGY FROM A SECONDARY SOURCE[23].

9B. ENERGY ANALYSIS OF TERTIARY WASTE RECYCLING SYSTEM THAT USES CYPRESS SWAMP INTERFACE ECOSYSTEMS IN FLORIDA[23].

conservation dollars are wisely spent when the investment ratio is moderately low.

ENERGY CONVERGENCE ON WETLANDS WITHIN A LANDSCAPE

In the same sense that energy converges in food chains, runoff waters from landscapes help converge and concentrate the energy embodied in sunlight, rain, wind, and substances from the uplifted land. The embodied energy is the energy required to generate the flow. These flows contribute to the formation of the geomorphologic characteristics of the wetland basin, to the maintenance of the existing ecosystem, and to the perpetuation of its roles in service to the landscape as a whole (e.g., aquifer recharge). Wetlands comprise about 10 to 20% of the landscape in Florida. The convergence effect allows them to develop high-quality structures, analogous to the high-quality organisms found at the tops of food chains, using the embodied energy of the larger total area in which they are embedded. The nutrient-trapping basin of the cypress dome is an example of such a structure. Because of the convergence effect, the swamp is the most valuable part of the landscape. This concept is illustrated in Figure 10.

When the capacity of the swamp to serve a recycling role for human society is considered, the value of the swamp may be considered to be 10 times greater than that of the general landscape. In systems that survive (both human and natural systems), structures with high embodied energy inputs generally are those that have important uses in that system. Economic benefits can result when these energy-expensive natural structures in natural systems can be tapped to serve human activities. If calculation of the investment ratio were to take into account the special role that wetlands play in the landscape, the investment ratio describing the use of swamps for sewage recycling would be even more favorable than the value calculated above.

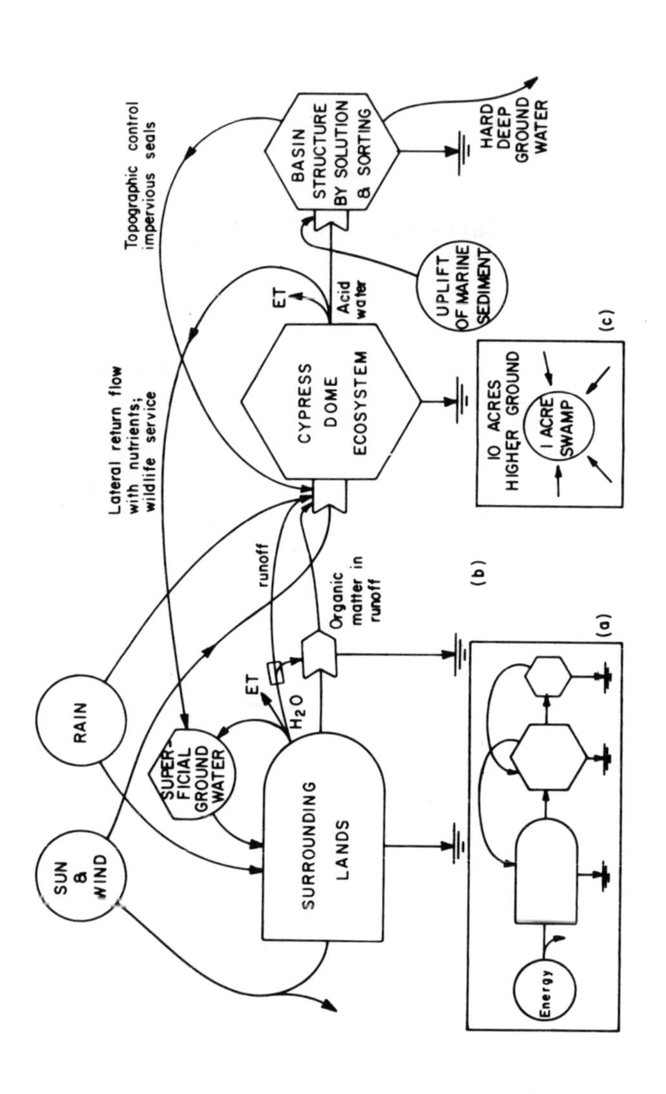

FIGURE 10. Convergence of Energy in a Landscape on a Cypress Dome Ecosystem

REFERENCES

1. Spangler, D.P., Gillespie, D., and Lundy, J. (1976) Stratigraphy and hydrologic correlation studies in several cypress domes in north-central Florida, pp. 110-128. Cypress wetlands for water management, recycling, and conservation. H.T. Odum and K.C. Ewel (eds.). Third Annual Report to the National Science Foundation (RANN) and The Rockefeller Foundation. Center for Wetlands, Univ. of Florida, Gainesville.

2. Duever, M.J., J.E. Carlson, L.A. Riopelle, and L.C. Duever (1978) Ecosystem analysis at Corkscrew Swamp. Cypress wetlands for water management, recycling, and conservation, pp. 534-570. H.T. Odum and K.C. Ewel (eds.). Fourth Annual Report to the National Science Foundation, Division of Applied Science and Research Applications, and The Rockefeller Foundation. Center for Wetlands, Univ. of Florida, Gainesville.

3. Ewel, K.C. and Mitsch, W.J. (1978) The effects of fire on species composition in cypress dome ecosystems. Fla. Sci. 41 (In Press).

4. Brown, S. (1978) A comparison of cypress ecosystems. Cypress wetlands for water management, recycling, and conservation, pp. 462-494. H.T. Odum and K.C. Ewel (eds.). Fourth Annual Report to the National Science Foundation, Division of Applied Science and Research Applications, and The Rockefeller Foundation. Center for Wetlands, Univ. of Florida, Gainesville.

5. Mitsch, W.J. and Ewel, K.C. Comparative biomass and growth of cypress in Florida wetlands. Amer. Midl. Nat. (In Press).

6. Ewel, K.C. (1978) Effect of fire and sewage on understory vegetation in cypress domes, pp. 92-103. Cypress wetlands for water management, recycling, and conservation. H.T. Odum and K.C. Ewel (eds.). Fourth Annual Report to the National Science Foundation, Division of Applied Science and Research Applications, and The Rockefeller Foundation. Center for Wetlands, Univ. of Florida, Gainesville.

7. Deghi, G.S. (1978) Growth rates of seedlings in four cypress domes. Cypress wetlands for water management, recycling, and conservation, pp. 319-340. H.T. Odum and K.C. Ewel (eds.). Fourth Annual Report to the National Science Foundation, Division of Applied Science and Research Applications, and The Rockefeller Foundation. Center for Wetlands, Univ. of Florida, Gainesville.

8. Straub, P. and Post, D.M. (1978) Rates of growth and nutrient concentration of trees in cypress domes, pp. 271-318. Cypress wetlands for water management, recycling, and conservation. H.T. Odum and K.C. Ewel (eds.). Fourth Annual Report to the National Science Foundation, Division of Applied Science and Research Applications, and The Rockefeller Foundation. Center for Wetlands, Univ. of Florida, Gainesville.

9. Deghi, G.S. (1977) Effect of sewage effluent application on phosphorus cycling in cypress domes. M.S. Thesis, Univ. of Florida. 143 pp.

10. Lugo, A.E., Nessel, J., and Hanlon, T. (1978) Studies on root biomass, phosphorus immobilization by roots and the influence of root distribution on plant survival in a north-central Florida cypress strand, pp. 802-826. Cypress wetlands for water management, recycling, and conservation. H.T. Odum and K.C. Ewel (eds.). Fourth Annual Report to the National Science Foundation, Division of Applied Science and Research Applications, and The Rockefeller Foundation. Center for Wetlands, Univ. of Florida, Gainesville.

11. Dierberg, F.E. and P.L Brezonik (1978) The effect of secondary sewage effluent on the surface water and groundwater quality of cypress domes, pp. 178-270. H.T. Odum and K.C. Ewel (eds.). Fourth Annual Report to the National Science Foundation, Division of Applied Science and Research Applications, and The Rockefeller Foundation. Center for Wetlands, Univ. of Florida, Gainesville.

12. Jetter, W. and Harris, L.D. (1977) The effects of perturbation on cypress dome animal communities, pp. 577-653. Cypress wetlands for water management, recycling, and conservation. H.T. Odum and K.C. Ewel (eds.). Third Annual Report to the National Science Foundation (RANN) and The Rockefeller Foundation. Center for Wetlands, Univ. of Florida, Gainesville.

13. McMahan, E.A. and Davis, L. (1978) Density and diversity of microarthropods in wastewater treated and untreated

cypress domes pp. 429-461. Cypress wetlands for water management, recycling, and conservation. H.T. Odum and K.C. Ewel (eds.). Fourth Annual Report to the National Science Foundation, Division of Applied Science and Research Applications, and The Rockefeller Foundation. Center for Wetlands, Univ. of Florida, Gainesville.

14. Davis, H. (1978) Effect of the treated effluent on mosquito populations and arbovirus activity. Cypress wetlands for water management, recycling, and conservation. H.T. Odum and K.C. Ewel (eds.). Fourth Annual Report to the National Science Foundation, Division of Applied Science and Research Applications, and The Rockefeller Foundation. Center for Wetlands, Univ. of Florida, Gainesville.

15. Allinson, J. and Fox, J.L. (1977) Coliform monitoring of waters associated with the cypress dome project, pp. 309-320. Cypress wetlands for water management, recycling, and conservation. H.T. Odum and K.C. Ewel (eds.). Third Annual Report to the National Science Foundation (RANN) and The Rockefeller Foundation. Center for Wetlands, Univ. of Florida, Gainesville.

16. Ordway, J.W. (1976) General site selection, design criteria, and costs for recycling wastewaters through cypress wetlands. Master's Paper. Univ. of Florida, Gainesville. 54 pp.

17. Fritz, W.R. and Helle, S.C. (1977) Tertiary treatment of wastewater using cypress wetlands. Boyle Engineering Corp., Orlando, Fla.

18. Heimburg, K.F. (1977) Hydrology and water budgets of cypress domes, pp. 56-67. Cypress wetlands for water management, recycling, and conservation. Third Annual Report to the National Science Foundation (RANN) and The Rockefeller Foundation. Center for Wetlands, Univ. of Florida, Gainesville.

19. Burns, L.A. (1978) Productivity, biomass and water relations of a Florida cypress forest. Ph.D. Dissertation. Dept. of Zoology, University of North Carolina, Chapel Hill. 213 pp.

20. Arkley, R.J. (1963) Relationships between plant growth and transpiration. Hilgardia 34:559-584.

21. McClurkin, D.C. (1965) Diameter growth and phenology of trees on sites with high water tables. U.S. For. Serv. Res. Note SO-22. 4 pp.

22. Mitsch, W.J. (1975) Systems analysis of nutrient disposal in cypress wetlands and lake ecosystems in Florida. Ph.D. Dissertation. Dept. of Environmental Engineering Sciences, Univ. of Florida, Gainesville. 421 pp.

23. Odum, H.T., Kylstra, C., Alexander, J., Sipe, N., Lem, P., Brown, M., Brown, S., Kemp, M., Sell, M., Mitsch, W., DeBellevue, E., Ballentine, T., Fontaine, T., Bayley, S., Zucchetto, J., Costanza, R., Gardner, G., Dolan, T., Boynton, W., and Young, D. (1976) Net energy analysis of alternatives for the United States, pp. 254-304. U.S. Energy Policy: Trends and Goals. Part V - Middle and Long-term Energy Policies and Alternatives. 94th Congress, 2nd Session Committee Print. Prepared for the Subcommittee on Energy and Power of the Committee on Interstate and Foreign Commerce of the U.S. House of Representatives. U.S. Govt. Printing Office, Washington, D.C.

10. BIOLOGICAL FLUIDIZED BEDS FOR NITROGEN CONTROL

John S. Jeris. Professor, Manhattan College and Vice President, Ecolotrol, Inc., Bethpage, New York

Roger W. Owens. Director R & D, Ecolotrol, Inc., Bethpage, New York

INTRODUCTION

With concern about (1) eutrophication caused by various nitrogenous species, (2) oxygen depletion in natural waters by ammonia and (3) the public health limitation of 10 mg/ℓ nitrate-nitrogen in drinking waters, an increased emphasis on the removal of nitrogen from wastewater has evolved.

Nitrogen is a relatively new component of wastewater that is considered necessary to remove in certain situations. The species of nitrogen that may require removal include organic nitrogen, ammonia, nitrite and nitrate. The process to be described will deal with the oxidation of ammonia to nitrate and the reduction of nitrite and nitrate to nitrogen gas, although it can be used for carbonaceous BOD removal as well.

The biological fluidized bed Hy-Flo* process is a novel high-rate system which is quite different than the typical suspended growth systems used. It has been used successfully for the treatment of wastewater at the Nassau County Bay Park, New York 60 MDG treatment plant for several years. Using three pilot plants of 36,000 to 72,000 gallons per day, BOD removal, nitrification and denitrification has been accomplished in about 45 minutes total detention time.

Early references to the use of fluidized granular beds include the works of Beer, Jeris, Mueller and Owens.[1 2 3] The biological fluid bed process developed, consists of passing wastewater upwardly through a reaction vessel which is partially filled with fine grained media as shown in Figure 1, at a velocity sufficient to impart "motion to" or "fluidize"

* A patented process of Ecolotrol, Inc., Bethpage, N.Y.

the bed. On the media surface, a biological mass grows as a firmly attached active mass which effectively consumes the waste as it passes by.

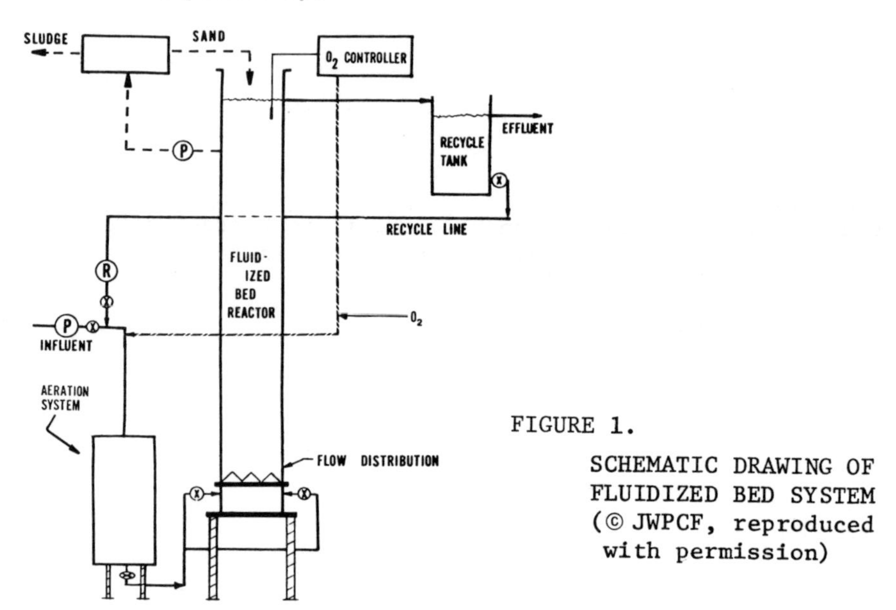

FIGURE 1.

SCHEMATIC DRAWING OF FLUIDIZED BED SYSTEM (© JWPCF, reproduced with permission)

From a biological point of view, the attached microorganisms may include any of the aerobic, facultative or anaerobic organisms typically found in trickling filter or suspended growth type of treatment systems. The predominating species would depend entirely on the waste contaminant being consumed and whether an aerobic or anaerobic environment is maintained, as well as other factors which affect biological growth. For example, if the contaminant were ammonia, a proper aerobic environment would allow nitrification to take place with the typical autotrophs, nitrosomonas and nitrobacter, present. If the contaminant were nitrate, the typical denitrifying organisms would grow if an anaerobic environment were maintained.

The medium acts as a support surface upon which the biological population thrives. Various fine grained media may be used to support the attached growth. To date, only activated carbon[1] and sand[2 3] have been used. Fluidizing the media allows solids to pass through and overcomes the problems associated with packed bed systems, such as high head losses, plugging, and frequent backwashing. Because small particles are used, a vast surface area is available for the growth of microorganisms. The fluid bed offers a large surface area for growth per unit of reactor volume. As all the media particles are in fluid motion, there is no contact between each particle, and intimate contact of the entire media surfaces with the waste stream is assured.

NITRIFICATION REACTORS

One nominal 72,000 gallon per day fluidized bed pilot reactor was operated for nitrification of secondary effluent. The reactor was two feet in diameter and fifteen feet high, having an empty bed detention time of approximately six minutes at the nominal flow rate. Constant flow conditions were used.

Figure 1 is a schematic drawing of the nitrification system. Flow is pumped through the pressurized oxygen mixing tank, labeled "aeration cone", where it is hydraulically mixed with pure oxygen gas. This aeration cone is designed so that the velocity of the wastewater entering the cone is higher than the rise rate of the bubbles and the velocity of the wastewater at the widest section of the cone is lower than the rise rate of the bubbles. This provides for a very long contact time of the oxygen gas with the wastewater.[4] Upon leaving the aeration cone, the wastewater flows through the fluidized bed reactor which contains sand. A flow distribution plate is located one foot above the bottom of the reactor, and the effluent port is located one foot below the top. Following the fluidized bed, the effluent flows into a sand return tank. Flow is introduced tangentially to this tank creating centrifugal action with aids in settling any growth covered sand particles contained in the effluent from the column. This tank is 30 inches in diameter and 36 inches high, with an overflow rate of 14,400 gallons per day per square foot (gpd/sf). The media which settles in this tank is pumped back into the fluidized bed through a port located 15 inches above the flow distribution plate. The effluent from the sand return tank enters a recycle tank from which part of the final effluent can be pumped back to the aeration cone. Liquid oxygen was used as the oxygen source for nitrification although air may also be used. For the effective use of oxygen, an oxygen probe may be used in conjunction with an oxygen controller.

Biological growth becomes evident by an increase in depth of the fluid bed. When the depth exceeds a prescribed level, the pump (Figure 1) is activated manually or automatically and pumps the biologically coated sand from the fluid bed to the Sweco vibrating screen. The sand is caught on top of the screen and the vibrating action moves it to the outer periphery of the circular screen to an opening which returns the sand to the fluid bed reactor. Biological growth sheared from the sand passes through the screen and becomes the excess sludge wasted.

NITRIFICATION RESULTS

As the influent wastewater to be nitrified was the effluent of the Bay Park wastewater treatment plant, the pilot plant results obtained represent a real situation. The Bay Park plant incorporates step-aeration activated sludge with primary and secondary clarification. Typical diurnal variations of flow and nitrogen occur. However, the pilot plant operated at constant flow most of the time and only the concentration variations were encountered.

a) Summer operation

Equilibrium data was obtained over a month's period of time on the fluidized bed and also on a pilot suspended growth reactor. Both pilot plants were provided with the same secondary treatment plant effluent and were run side by side. The suspended growth reactor had a 12 hour detention time while the fluid bed had only 12 minutes. The daily influent and effluent NH_3-N results are given in Figure 2. The slight difference in influent NH_3-N concentration is due to the different times used in collecting the samples. It is quite evident that both systems can readily oxidize the ammonia and achieve virtually 100 percent removal. The two days on which greater than 1 mg/ℓ ammonia-nitrogen was present in the effluent from the fluid bed, insufficient oxygen was provided as an automatic oxygen controller was not available.

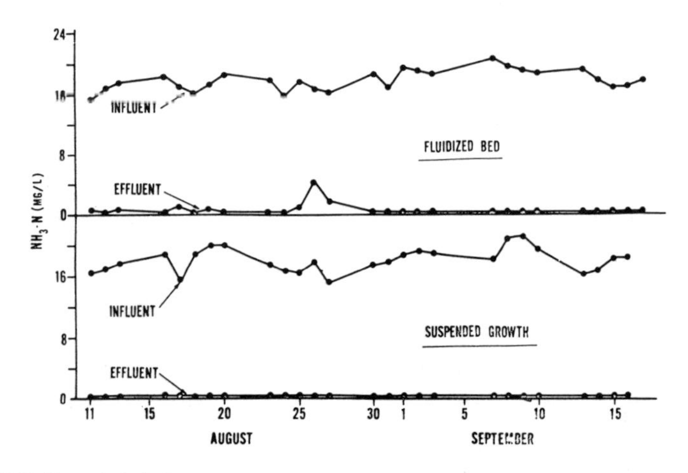

FIGURE 2. DAILY AMMONIA INFLUENT AND EFFLUENT RESULTS

Figure 3 shows the effluent suspended solids data for the two systems. As might be expected, the effluent from the suspended growth system gave erratic results as high suspended solids in excess of 30 mg/ℓ, were evidenced several times during the month. The light fluffy nature of the nitrifying activated sludge was the cause of these results. Since the fluid bed, is in effect a fixed film biological treatment system, an effluent solids problem was not encountered and the average concentration was less than 15 mg/ℓ suspended solids.

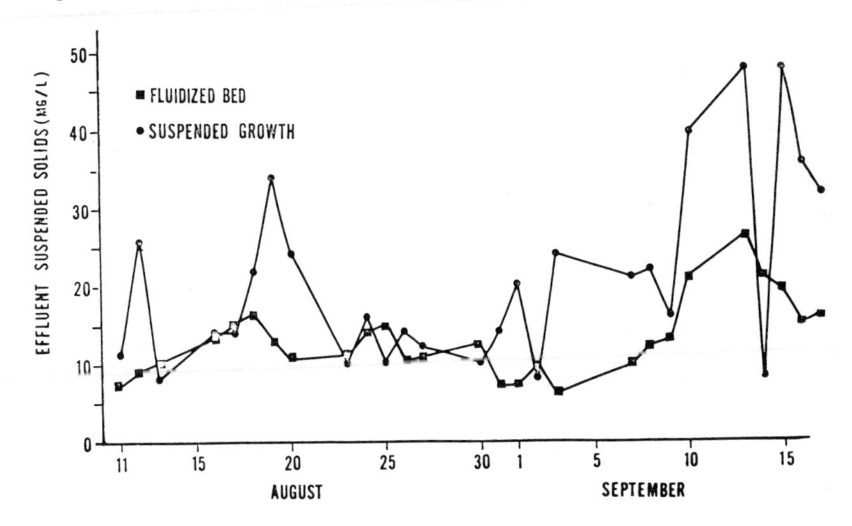

FIGURE 3. COMPARISON OF EFFLUENT SUSPENDED SOLIDS FOR FLUIDIZED BED AND SUSPENDED GROWTH SYSTEMS

Very little biological growth or bed height increase was obtained during summer operation, as the high number of samples taken for analysis probably accounted for the daily sludge production. The MLVSS concentration varied somewhat with bed height but averaged about 8500 mg/ℓ. When the sampling of particles from the fluid bed was discontinued, the suspended solids were measured. Samples of the influent and effluent suspended solids averaged over a six-day period were 9 and 13 mg/ℓ respectively or a 4 mg/ℓ increase in suspended solids.

A typical ammonia profile through the fluid bed reactor is given in Figure 4. It can be seen that a reactor of about 40 inches leaves only one mg/ℓ ammonia nitrogen in the effluent and at 75 inches virtually all the nitrogen is gone. It should be noted that recycle is employed in this system and the low concentration of 6 mg/ℓ shown as the starting substrate concentration represents a dilution with the recycle.

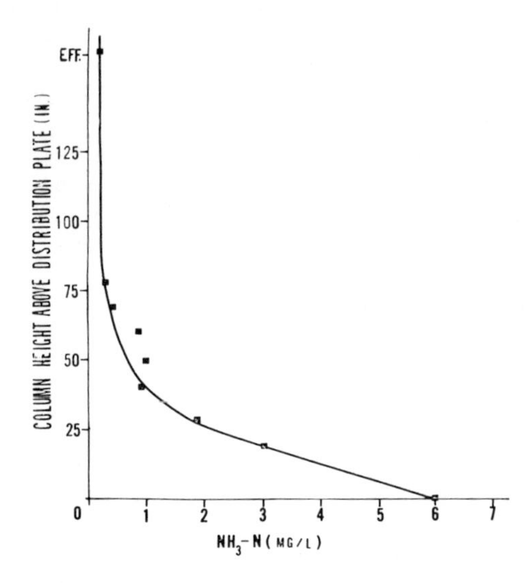

FIGURE 4. AMMONIA PROFILE THROUGH NITRIFICATION REACTOR (Copywrited JWPCF, reproduced with permission)

Oxygen utilization data collected during this period was analyzed in conjunction with the data for NH_3-N and COD removal and the following equation was developed based on a regression analysis:

Oxygen Consumed = 4.3(NH_3-N oxidized) + 0.9 (COD removed).

As pH affects the rate of nitrification, a study was made to determine its effect by operating with and without alkalinity addition to the influent wastewater. Table 1 shows the effects at a wastewater temperature of 24°C. A drop in pH from 6.1 to 5.8 resulted in a decline in efficiency, from 99% to 87.4% or an increase in the effluent NH_3-N concentration of 2.5 mg/ℓ. This decline in pH caused a decrease of 20 mg/ℓ expressed as $CaCO_3$. An effluent alkalinity of 50 mg/ℓ seems adequate if a very high efficiency of ammonia removal is to be obtained.

(b) Winter operation

Cold wastewater temperatures extending between 13° and 16°C were studied. Unfortunately the organic loading was extremely high due to upset conditions at the Bay Park plant which was providing the secondary effluent for nitrification. The average influent COD was reduced from 109 to 45 mg/ℓ while greater than 99 percent ammonia removal was obtained. The data is presented graphically in Figure 5 and is summar-

TABLE 1. Effect of pH and Alkalinity on Nitrification

WITHOUT ALKALINITY ADDITION

	Effluent		NH_3-N mg/ℓ	
	pH	Alk(mg/ℓ)	Influent	Effluent
	5.7	20	21.6	2.6
	6.1	40	19.6	1.0
	6.2	52	20.1	2.1
	5.9	32	21.0	4.0
	5.9	30	20.4	3.7
	5.7	22	19.6	2.1
	5.8	28	22.1	4.1
	6.0	39	21.8	2.2
	5.8	26	21.3	2.0
	5.7	22	21.6	2.1
	5.4	12	21.0	2.0
	5.4	12	24.4	3.5
	5.5	11	23.5	4.1
Average	5.8	27	21.4	2.7

WITH ALKALINITY ADDITION

	Effluent		NH_3-N mg/ℓ	
	pH	Alk(mg/ℓ)	Influent	Effluent
	6.1	44	18.7	0.4
	5.9	26	19.0	0.8
	6.5	86	17.9	0.07
	6.3	46	19.3	0.3
	6.3	62	18.5	0.1
	6.3	42	19.9	0.4
	6.0	50	18.2	0.0
	6.1	46	18.8	0.4
	6.2	52	19.3	0.5
	6.1	36	21.0	0.4
	6.2	42	16.0	0.0
	6.1	46	18.5	0.0
	6.3	52	18.5	0.0
	6.1	50	18.5	0.0
	6.1	48	19.6	0.0
	6.1	44	20.4	0.7
	6.2	52	18.8	0.0
	6.0	44	20.2	0.0
	6.1	44	19.9	0.1
	5.9	40	21.0	0.0
	5.9	38	19.3	0.3
Average	6.1	47	19.1	0.2

ized in Table 2. It should be noted that only 26 minutes were required to obtain 99% ammonia oxidation along with a reduction of 64 mg/ℓ COD. These results were obtained with 12,000 mg/ℓ MLVSS present in the fluidized bed reactor.

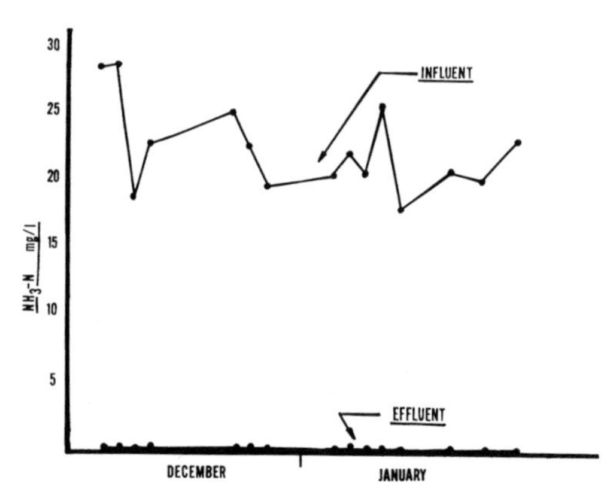

FIGURE 5. AMMONIA REMOVAL UNDER COLD WEATHER CONDITIONS

TABLE 2. Average Nitrification Results

Inflow (gpm)	4.7
Recycle (gpm)	40.0
Detention time (min)	26
Temperature °C	14.5
Influent NH_3-N(mg/ℓ)	21.9
Effluent NH_3-N(mg/ℓ)	0.1
Influent COD (mg/ℓ)	109
Effluent COD (mg/ℓ)	45
Influent pH	7.3
Effluent pH	6.3
MLVSS (mg/ℓ)	12,000

Additional winter data was also collected during a period when the fluid bed reactor was removing both primary effluent BOD and nitrogen simultaneously. The primary effluent BOD entering the reactor was 128 mg/ℓ and the effluent averaged 23 mg/ℓ. In addition approximately 80% of the ammonia and 60% of the organic nitrogen was removed. During this period the effluent pH was consistently less than 6.0 because of the low alkalinity present in the wastewater. This inhibited complete nitrification of the wastewater. Temperatures during the study extended from 11 to 13.5°C with an average of 12.5°C. The average MLVSS were 20,000 mg/ℓ with a F/M ratio of 0.12 and a solids retention time of 15.4 days.

DENITRIFICATION

The denitrification reactor is constructed of 0.25 inch thick plexiglass and is 1.5 feet in diameter by 15.5 feet high. With the sand medium covered with growth, the height of the fluidized bed in operation is about 12 feet. Based on the height of the fluidized bed, the empty bed detention time is 6.5 minutes at the design flow of 36,000 gpd or 15 gpm/sf of surface area. Methanol was used as the carbon source during this work.

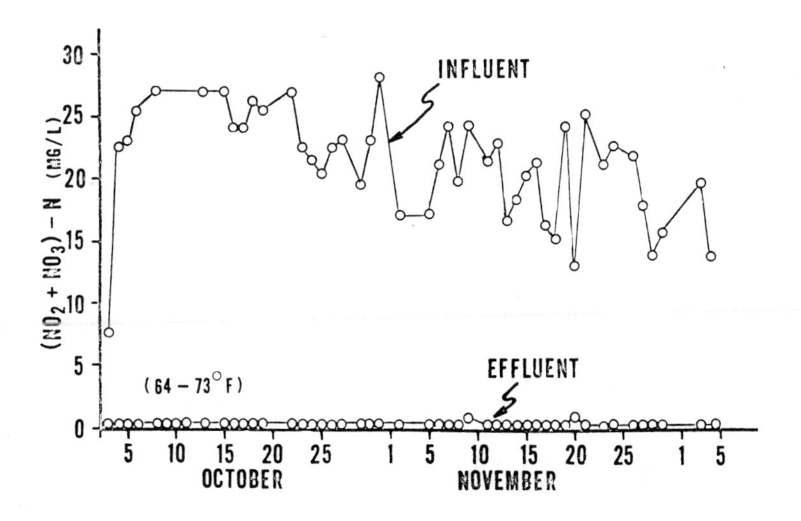

FIGURE 6. DAILY DENITRIFICATION RESULTS

In Figure 6 are the daily results of the influent and effluent NO_2-N and NO_3-N concentrations. Removals of 99% were obtained throughout this portion of the study. A summary of these results is also given in Table 3. The temperature range was 18°-23°C and no decrease in efficiency occurred because of a pH which declined to 5.8 and a dissolved oxygen range of 0-4 mg/ℓ. Although the methanol to nitrate ratio averaged 4.2, a ratio close to 3.0 would probably have sufficed if automated controllers for methanol had been available.

TABLE 3. Summary of Operation

Flow (gpm)	26.0
Flux (gpm/sf)	15.6
Empty Bed Detention Time (min)	6.5
Temperature (°F)	68
Influent pH	6.6
Effluent pH	7.5
Influent DO (mg/ℓ)	1.5
MeOH/NO_3-N	4.2
Influent (NO_3 + NO_2) - N(mg/ℓ)	21.5
Effluent (NO_3 + NO_2) - N(mg/ℓ)	0.2
Percent (NO_3 + NO_2) - N Removed	99.0

The carbon source used throughout most of the test period was industrial grade methanol. Methanol was used primarily because it produces a low cell yield, is very soluble in water and economical. Previous studies have shown that a three to one ratio of methanol to nitrogen is required. Results of the "methanol limiting" studies verify this ratio as being fairly accurate. Figure 7 shows the efficiencies obtained at various ratios.

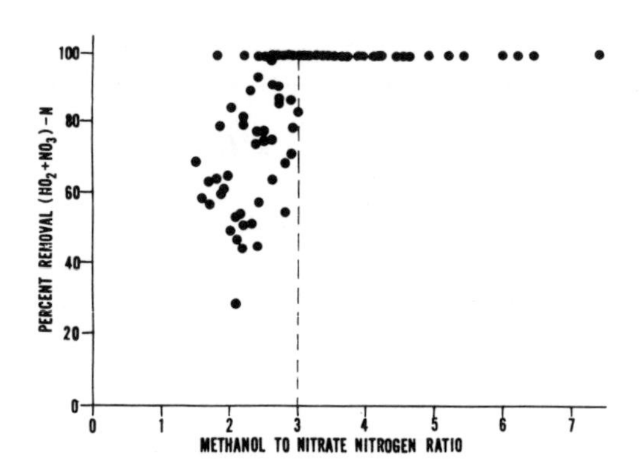

FIGURE 7. EFFECT OF METHANOL TO NITROGEN RATIO ON REMOVAL EFFICIENCY
(Copywrited JWPCF, reproduced with permission)

Figure 8 shows a typical nitrogen profile through the column. From this curve, it can be seen that essentially all of the influent nitrogen is removed after a detention time of one minute in the column and that the reactor is overdesigned.

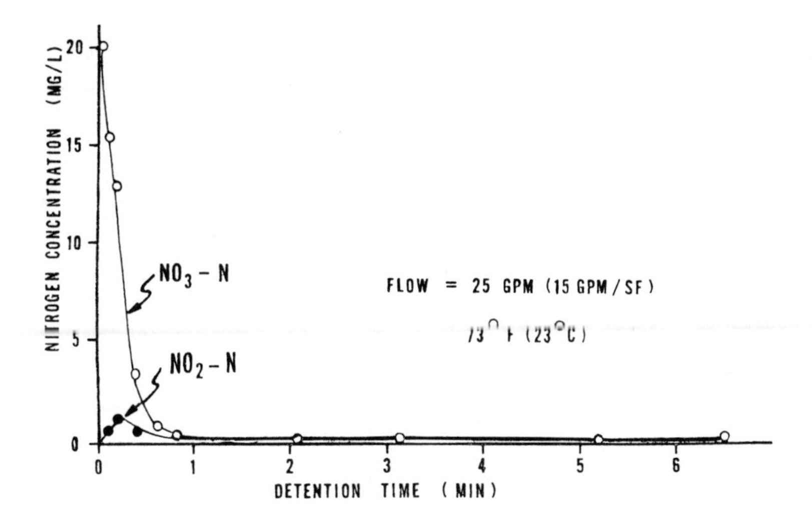

FIGURE 8. PROFILE OF NITRITE AND NITRATE THROUGH REACTOR (Copywrited JWPCF, reproduced with permission)

Temperature has a significant effect on biological growth. Obviously, as the temperature decreases, so does the reaction rate of the denitrifying organisms. Figure 9 shows the design loading rates obtained at various operating temperatures during 18 months of continuous operation. The loading factor is an order of magnitude greater than that obtainable by conventional suspended growth technology.[5] This curve can be used for design purposes by selecting the design temperature and obtaining the reactor nitrogen loading rate.

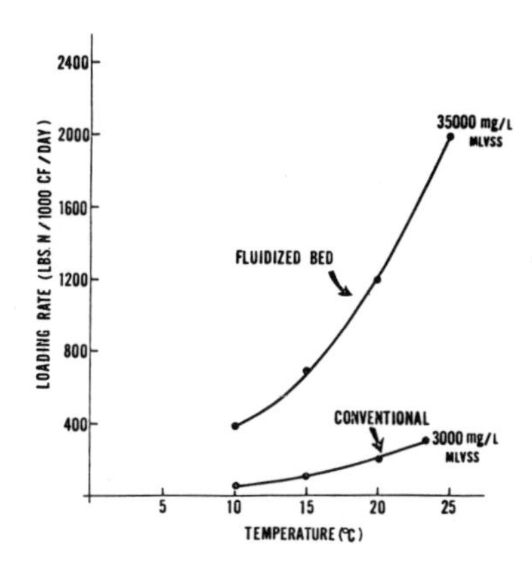

FIGURE 9. EFFECT OF TEMPERATURE ON NITROGEN LOAD (Copywrited JWPCF, reproduced with permission)

For a period of one week, flow was increased to 40 gpm or 24 gpm/sf, giving an empty bed detention time of approximately 4 minutes with no detrimental effect on the effluent quality. Diurnal variation in flow from 20 to 40 gpm also produce no change in effluent quality. To determine the effect of power failure on the system, the reactor was shut down for a period of 17 hours and repeated at a later date for a period of 25 hours. Within 45 minutes and 15 minutes respectively, after start-up, the system was again achieving 99% removal. During a high nitrogen concentration study,[6] which would simulate a nitrate containing industrial waste, the influent oxidized nitrogen averaged 55 mg/ℓ and the effluent 2.5 mg/ℓ for an overall removal of 95%. On two days of this study, however, insufficient methanol was fed to the reactor. If these two days are excluded from the average, the removal of nitrogen was 99%. Operation during the winter months with temperatures as low as 8 to 12°C did not significantly affect performance.

DISCUSSION

The results obtained demonstrate the feasibility of using fluidized bed systems for nitrification and denitrification of municipal wastewater. Under winter conditions it was shown that 99% nitrification and denitrification can be achieved in 26 and 6.5 minutes respectively.

What the fluidized bed actually does is combine the best features of activated sludge and trickling filtration into one process. Offering a fixed film and a large surface area, fluidized bed systems offer the stability and ease of operation of the trickling filter as well as the greater operating efficiency of the activated sludge process. More importantly, treatment is accomplished in significantly less space and time which can be translated into less cost than conventional treatment. The primary reason for this savings in space, cost and treatment time is that the concentration of active biomass (volatile solids) in the fluidized bed system is in the order of 8,000 to 40,000 mg/ℓ which is far greater than conventional treatment systems. Table 4 summarizes mixed liquor volatile suspended solids (MLVSS) concentrations of various suspended growth treatment processes showing the marked differences in biological mass.

TABLE 4. Comparative MLVSS Concentrations

Treatment System	MLVSS in mg/ℓ
Pure oxygen activated sludge	3,000 - 5,000
Conventional activated sludge	2,000 - 3,000
Suspended growth nitrification	1,000 - 1,500
Suspended growth denitrification	1,500 - 2,500
Fluid Bed	
BOD removal	15,000 -20,000
Nitrification	8,000 -12,000
Denitrification	30,000 -40,000

Comparing surface area per unit volume of reactor for fixed film systems in Table 5, it is again very evident that the fluid bed system has a far greater surface area available for biological growth than eigher trickling filters or rotating biological contactors.

TABLE 5. Comparative Media Surace Areas

Treatment System	Surface Area sf/cf of reactor volume
Trickling filter	12 - 30
Rotating Biological contactors	40 - 50
Fluidized bed	800 - 1,200

Being a biological system, growth of the particle must be controlled. It has been found that by pumping a portion of the fluid bed through a vibrating screen, and returning the sand retained on the screen to the fluid bed offers excellent control. The excess solids passing the screen have a concentration of 1 to 2 percent and can be further processed by a number of solids handling methods. Operation of the vibrating screen for solids control has been required for only 10 to 20 minutes each day and the volume of solids needing further handling is about 0.2 to 0.3% of the wastewater flow, thus simplifying solids handling substantially.

This technique for solids handling eliminates the need for a secondary clarifier as results continuously show suspended solids in the effluent to be very low when using the vibrating screen. The economic savings brought about by eliminating the clarifier are substantial.

INSTALLATIONS AND ECONOMICS

Treatment systems for denitrification have been installed at a nursing home (18,000 gpd) and a small development (250,000 gpd). A larger system is now under construction and scheduled for completion in early 1979 at Pensacola, Florida. This 24 MGD fluid bed system is part of an advanced waste treatment facility requiring carbon, nitrogen, phosphorus and solids removal. It incorporates four 19' x 19' denitrification reactors. A 90% reduction in land requirements was realized over the comparable suspended growth systems. Since pilings were required at this site, the great reduction in area resulting from the use of the fluidized bed saved the community about $1.5 million in piling and site preparation costs. Further savings due to reduction in concrete costs were also realized.

A cost comparison for a 16 MGD treatment plant for denitrification was made for the suspended growth and the biological fluidized bed systems. The comparison given in Table 6 shows a capital cost saving of over $2,000,000. From an energy point of view the costs are about the same.

TABLE 6. COST COMPARISON

	Suspended Growth	Fluid Bed
Tankage	$1,100,000	$ 310,000
Pumps	50,000	370,000
Wet Wells	50,000	120,000
Clarifiers	2,000,000	-
Solids separator	-	60,000
Capital Cost	$3,200,000	$ 860,000
Annual Energy Cost	$ 28,000	$ 26,700

An innovative biological process has been described which relies on well established biological principles and offers the easily understood concepts of high surface area and high concentrations of biological mass. The savings of money and land with its use are very significant. Application should also be considered in existing treatment plants with little room for expansion.

In effect this process has solved the problem of concentrating the biological mass in a treatment reactor to a far greater degree than has been done by any other biological treatment process. Basically this is translated to mean smaller treatment plants for any flow and a major reduction in cost.

REFERENCES

1. Jeris, J.S., Beer, C., and Mueller, J.A. "High Rate Biological Denitrification Using a Granular Fluidized Bed." JWPCF, 46, 2118-2128 (1974).

2. Jeris, J.S. and Owens, R.W.. "Pilot Plant, High-Rate Biological Denitrification." JWPCF, 47, 2043-2057, (1975).

3. Jeris, J.S., Owens, R.W., Hickey, R. and Flood, F. "Biological Fluidized-bed Treatment for BOD and Nitrogen Removal." JWPCF, 49, 816-831 (1977).

4. Speece, R.E. et al. "Downflow Bubble Contact Aeration." Proc. 25th Industrial Waste Conf., Purdue Univ. (1970).

5. "Nitrification and Denitrification Facilities Wastewater Facilities." EPA Technology Transfer Seminar Publication. 1973.

6. Gasser, R.F., Owens, R.W. and Jeris, J.S. "Nitrate Removal from Wastewater Using Fluid Bed Technology." 30th Purdue University Industrial Waste Conference, Ann Arbor Science (1975).

11. PROCESSING BIOLOGICAL NUTRIENT REMOVAL SLUDGES

David Di Gregorio, Director of Sanitary Engineering Technology, Eimco PMD, Envirotech Corporation, Salt Lake City, Utah

ABSTRACT

The development of an appropriate sludge processing scheme for a wastewater treatment facility is influenced by the quantities and characteristics of the sludge to be processed, and the method of ultimate disposal of the sludge to the environment. Methods for sludge processing and disposal are examined within the context of biological nutrient removal systems.

INTRODUCTION

Two of the points of consideration in the development of an appropriate sludge processing conceptual design for a wastewater treatment facility are: (1) knowledge of the quantities and physical and chemical characteristics of the sludge to be processed and (2) the method of ultimate disposal or disposition of the sludge to the environment.

Knowledge of the quantities and characteristics of the sludge to be processed is important in terms of evaluating the appropriateness of the various sludge processing alternatives which exist. This knowledge, of course, could also impact on the method chosen for ultimate disposal.

Selection of appropriate processing alternatives is also affected by the final disposition of the sludge. For example, disposal by land application would require a consideration of the quantities and characteristics of the sludge, and the availability and capacity of the land disposal sites(s). Different sludge processing alternatives would be indicated for situations involving abundant land disposal opportunities as compared to limited availability and/or capacity of land disposal sites. Additionally, the type of land application method (i.e., cropland versus sanitary landfill) would also impact on the sludge processing scheme selected.

Biological nutrient removal systems are relatively new technologies. The waste sludges produced from these systems, however, are not significantly different from those produced from more conventional treatment systems. The purpose of this paper will be to discuss sludge processing unit operations and systems within the context of disposal of biological nutrient removal sludges.

SLUDGE DISPOSAL ALTERNATIVES

The first point of consideration in the development of an appropriate sludge processing scheme is the method of ultimate disposal. There are several alternatives available. Ultimately, the excess sludges or residues produced at a waste treatment facility must be disposed of by one of four means. These are:

(1) Croplands for soil conditioning or fertilizer value;
(2) Marginal lands for land reclamation purposes;
(3) Sanitary landfill without land reclamation purposes; and
(4) Ocean disposal.

Consideration to state and federal regulations, land availability, soil characteristics, geology, climate, sludge characteristics, and cost are necessary to select the appropriate ultimate disposal alternative for a specific location.

Once an ultimate disposal method has been selected, it is necessary to select the sludge processing unit operations which will render the sludge fit for the ultimate disposal method. Sludge preparation for the four disposal methods listed above may involve substantially different sludge processing operations especially when regulations, land characteristic and sludge property interactions are considered. Table I presents a brief summary of the considerations necessary for the selection of appropriate sludge processing operations.

Table I. Sludge Processing Considerations

Ultimate Disposal Method	Sludge Form	Constraints
Croplands for soil conditioning or fertilizer value	Liquid or cake	Regulations Land Availability
Marginal lands for land reclamation purposes	Liquid, cake or ash	Geology Climate
Sanitary landfill without land reclamation purposes	Cake or ash	Sludge characteristics Cost

As shown in Table I, the ultimate disposal method determines the extent to which the waste sludge must be processed. Further, constraints, such as those listed in Table I, impact heavily on the selection of specific unit operations. For example, if cropland and marginal land were unavailable and if the sanitary landfill capacity were limited, processing sludge to an ash residue would be necessary.

SLUDGE PROCESSING ALTERNATIVES

Sludge processing operations may be classified in five categories:

(1) Thickening
(2) Stabilizing
(3) Conditioning
(4) Dewatering
(5) Reduction

There are several unit process alternatives for each category resulting in a multitude of sludge processing system alternatives. Selection of the appropriate sludge processing system alternative is determined after consideration of sludge characteristics, methodology associated with disposal of the sludge and cost.

Each of the above categories will be examined keeping within the context of biological nutrient removal flow sheets.

Thickening

The purpose of sludge thickening is to reduce the volume of sludge to be disposed of or handled by subsequent unit operations. The three most prominent thickening methods are gravity, flotation and centrifugal. High costs and maintenance have contributed to a limited use of centrifuges for thickening and thus only gravity and flotation will be discussed here.

(1) Gravity thickening. Gravity thickening is the simplest and least expensive of the three prominent thickening alternatives. As the performance information in Table II indicates, gravity thickening is effective for primary and bio filter sludge applications. Its performance is relatively poor, however, for waste activated sludge applications.

Table II. Expected Gravity Thickener Performance for Biological Sludges

Type of Sludge	Loading lb/day-ft^2	Underflow wt. % TS
Primary	20	8-10
Bio Filter	8	5-7
Activated	4-5	1.5-2.0

The most serious problem encountered with gravity thickening is the tendency for sludges to go septic while maintained in the thickener, especially during warmer periods. Septicity causes odors, floating sludge, poor solids capture and diminished sludge dewaterability. Septicity can be controlled somewhat by chlorine addition or maintaining a solids retention time in the thickener of less than 24 hours.

Sludge septicity or periods of anaerobic conditions within gravity thickeners would be of particular concern when applied to biological phosphorus removal systems such as the Bardenpho process. Anaerobic conditions within the thickener could result in phosphorus being stripped from the biological solids and recycled back to the main plant treatment processes. A build-up of phosphorus within those processes would occur and result in deteriorated phosphorus removal efficiency.

(2) Flotation thickening. Dissolved air flotation (DAF) is the most commonly used method for thickening waste activated sludges. It is seldom employed for primary or bio filter sludge applications since these sludges thicken readily by gravity.

Figure 1 presents DAF thickening performance results for a variety of biological sludges. These data imply that the waste sludges produced from biological nutrient removal flow sheets can be expected to thicken to concentrations similar to more conventional biological sludges. At normal air: solids ratios (2-3%), float solids of 3-4 weight percent total solids can be expected.

Solids capture of at least 95% can be obtained without polymer flocculation from properly sized and operated DAF units. Loadings of less than 2 lb/hr-sq ft and 1.5 gal/min-sq ft not only insure excellent solids capture but also provide concentrated float. Polymer flocculation may be employed and may improve solids capture and float solids, especially at high loading conditions.

Sludge Stabilization

The purpose of sludge stabilization is primarily to render sludges less putrescible and therefore fit for land disposal. Digestion, both aerobic and anaerobic, lime stabilization and heat treatment are the principle processes of stabilization.

(1) Anaerobic digestion. Anaerobic digestion is one of the most commonly used stabilization processes today. Many states require that sludges,

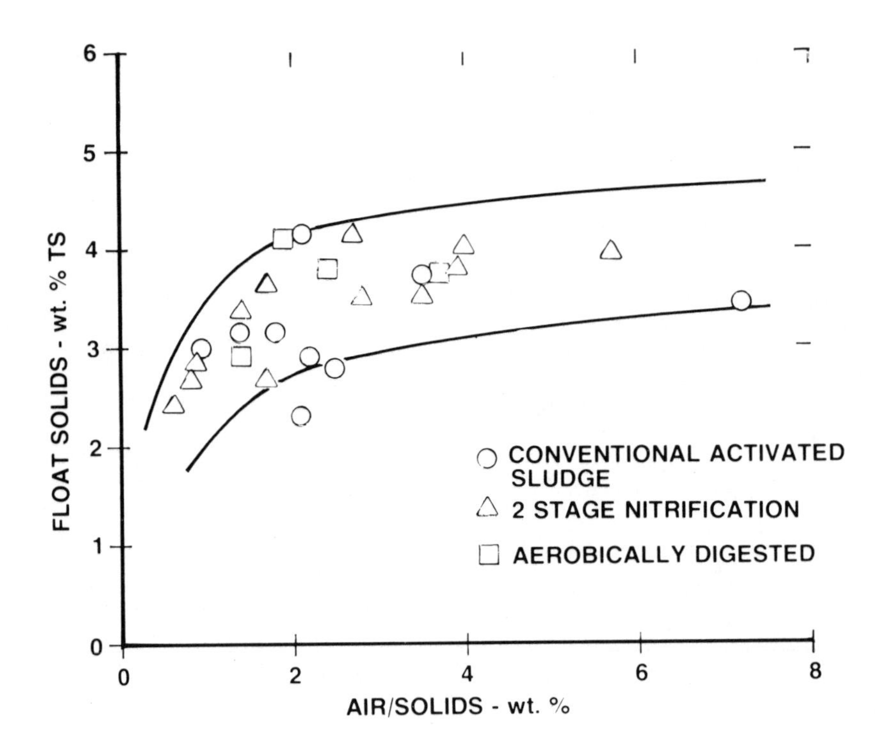

FIGURE 1. FLOTATION THICKENER PERFORMANCE FOR VARIOUS WASTE BIOLOGICAL SLUDGES

especially raw primary sludges, be digested prior to being disposed of on land.

There are several benefits derived from the anaerobic digestion process. Sludges, particularly raw primary sludges, are rendered less obnoxious during digestion. Pathogenic bacteria populations and organic matter undergo substantial reductions. Volatile solids destructions of 40-50 percent by weight are common, thereby resulting in reductions in the quantity of sludge to be disposed of. The organic matter destroyed during anaerobic digestion is converted to gaseous end products, primarily methane and carbon dioxide. The methane is of special interest since it is a valuable energy source which is often used for heat and power generation within treatment plants. The increased popularity of anaerobic digestion in recent years has largely been due to treatment plant operating cost economies which can be realized with proper use of methane as an energy source.

Digester supernatant quality merits special attention in connection with the biological nutrient removal application. Digester supernatant is relatively poor in quality in terms of suspended solids, BOD, nitrogen and phosphorus. The recycle load on the aeration system must be considered when applying anaerobic digestion for sludge stabilization. Recycle of supernatants produced from anaerobic digestion of waste sludge produced from biological nitrification systems results in a nitrogen and phosphorus recycle of approximately 15-25% based on raw wastewater nutrient load. However, a phosphorus recycle of 40% can occur for treatment systems designed for biological phosphorus removal. Separate chemical treatment or disposal of the digester supernatant is desirable in this situation.

High rate anaerobic digestion processes are generally designed for a solids retention time of 15-20 days. Volatile solids destruction of 40-50% is typical. End products of aerobic digestion are carbon dioxide, water, cell debris and non-biodegradable materials.

(2) Aerobic digestion. The principal advantages of aerobic digestion over anaerobic digestion are lower capital costs, better quality supernatant and less operational complexity. Comparable levels

of volatile solids destruction are achieved at shorter detention times than anaerobic digestion. The main disadvantages of aerobic digestion are higher operating costs and the inability of the process to produce methane, a useful energy source produced with anaerobic digestion.

The primary purpose of aerobic digestion is to stabilize raw and waste biological sludges to produce an end product suitable for subsequent processing or disposal. Stabilization is achieved with prolonged aeration under food-limiting conditions. During aerobic digestion, aerobic and facultative bacteria utilize cell protoplasm as a food source. Under prolonged aeration, old cells die releasing biodegradable materials for use by other bacteria.

Near maximum levels of volatile solids destruction are generally achieved within a solids retention time of 15 days. Volatile solids destruction of 40-50% is typical. Longer digestion periods have been used, especially when primary sludge is fed to the digester.

Aerobic digestion of excess sludges from biological nutrient removal systems is unnecessary since these sludges are already considered to be aerobically digested. These sludges normally have sludge ages of 20-30 days or more, and volatile contents of 50-60%. For biological nutrient removal systems designed with primary clarification, aerobic digestion of primary sludge may be practiced for stabilization.

(3) Heat treatment. Heat treatment is a process which can serve both sludge stabilizing and conditioning needs. Sludge is heated to temperatures of 350-450°F and maintained at 150-250 psig for detention times of 20-30 minutes. Cooking sludge in this manner results in complete sterilization and a significant improvement in the drainage characteristics of the sludge.

Heat treatment results in solubilization of a portion of the sludge. Waste biological sludges are solubilized to a greater extent than primary sludge. The solubilized solids report to the liquid fraction of the heat treated sludge resulting in a dark brown liquor high in BOD, ammonia and phosphorus. When this liquor is recycled to the main treatment works, the BOD, ammonia and phosphorus load to the biological system is generally increased by 15-20%.

The recycle nutrient load on the aeration system of biological nutrient removal systems must be considered when stabilizing waste sludge by heat treatment. This is especially true for biological nutrient removal systems designed without primary clarifiers since biological sludge tends to solubilize to a greater extent than primary sludge. As much as 40-60% of the nitrogen and phosphorus content of waste biological sludges can be "solubilized" during heat treatment. Recycle of heat treatment liquor results in much the same nitrogen and phosphorus load increases to the main treatment works as discussed above for anaerobic digestion.

(4) Lime stabilization. Waste sludges may be stabilized by the addition of lime in quantities sufficient to maintain a pH above 11. Substantial reductions in pathogenic bacteria occur.

The quantity of lime required to effectively stabilize sludge is dependent on the type of sludge and the length of time the pH is maintained above 11. Maintaining sludge pH for a two week period is generally considered necessary to insure effective stabilization. A lime dose of 10-20 weight percent is required for stabilization of primary sludge, and a 30-50 weight percent lime dose is required for waste activated sludge.

The advantages of lime stabilization are that it is a simple process requiring little capital investment. Operating costs range from approximately 10 $/TDS for primary sludge to 20-25% $/TDS for waste activated sludge. The disadvantages of lime stabilization are that no organic matter destruction occurs, and the quantity of sludge to be disposed of increases due to the addition of lime solids.

Conditioning

The purpose of sludge conditioning is to improve the thickening and/or dewatering properties of sludge. Chemical conditioning and heat treatment are the two most prominent sludge conditioning methods.

(1) Chemical conditioning. Thickening and dewatering properties of sludge are improved with

chemical conditioning. Polymers, ferric chloride and lime are the most popular types of chemical conditioners used. The type or combination of chemical conditioners and the quantity necessary to effectively condition sludge is dependent on its physical and chemical characteristics.

Gravity and flotation thickening processes are normally operated without chemical conditioning. For those applications where chemical conditioning is desirable, cationic polymers are employed. Conditioning may improve allowable solids loading, solids capture and thickened sludge concentration.

Nearly all of the waste sludges produced from wastewater treatment require some form of chemical conditioning to maximize the performance of sludge dewatering operations. Most dewatering processes will not function without conditioning. Polymer, ferric chloride and lime either individually or in combination are the most prominent chemical conditioners employed. The preferred chemical conditioning mode is dependent on the type of dewatering process and the type of sludge. A more detailed discussion of chemical conditioning effectiveness and costs is contained in connection with the discussion of sludge dewatering processes.

(2) Heat treatment. Heat treatment of sludges accomplishes not only stabilization, but conditioning. Heat treating sludges substantially improved the thickening and dewatering characteristics of sludges. Sludges are generally more effectively conditioned by heat treatment than by chemical conditioning. The effectiveness of heat treatment for sludge conditioning is illustrated in connection with the discussion of sludge dewatering processes. The advantages and disadvantages of heat treatment for stabilization apply to conditioning also.

Dewatering

The purpose of sludge dewatering is to remove sufficient water from liquid sludge so as to produce a cake with the physical handling characteristics and/or solids content optimal for subsequent processing or disposal. Sludge dewatering is accomplished with the use of any one of several dewatering devices. The dewatering approach selected is determined from an examination of sludge characteristics and the method of ultimate disposal of the sludge to the environment.

(1) Sand drying beds. Drying on sand beds is one of the most commonly used methods of sludge dewatering. This method is normally appropriate only for aerobically or anaerobically digested sludges. Performance of sand drying beds is affected by climate, sludge characteristics and operating procedures (i.e., drying time, sludge depth and use of conditioning aids). Most sand drying bed installations are operated to produce a cake of 10-15% total solids. Extended drying periods of 2-3 months and favorable climate, however, have resulted in cake solids as high as 40-45%. Chemical conditioning using polymers has also improved bed performance by reducing required drying time and producing solids contents of 10-30%.

(2) Vacuum filtration. Vacuum filtration has been the most widely used method for mechanically dewatering sludges. At the present time, there are over 2500 municipal vacuum filter installations in the U.S. Vacuum filtration has enjoyed widespread usage since it can be applied to virtually any plant size and to a wide variety of sludge including waste biological sludge.

Table III lists expected vacuum filtration performance for various sludges produced from biological nutrient removal systems. A review of the information presented in Table III reveals that the vacuum filtration characteristics of sludge are generally diminished by biological stabilization. Filtration characteristics are improved by stabilization by heat treatment. From the limited information available in the literature, it appears that lime stabilization of sludges generally improves filtration performance.

(3) Centrifugation. Centrifugation has been a popular method for dewatering waste sludges. The advantages of centrifugation over other mechanical dewatering alternatives are that it is a simple, compact, totally enclosed operation. The main disadvantages have been high maintenance and chemical costs and low cake solids at reasonable solids recovery levels.

Table III. Expected Vacuum Filter Performance

Sludge Type	Chemical Conditioning, wt. %		Filter Yield (a)	Cake Solids, wt. % TS	
	Polymer (b)	$FeCl_3$ + CaO	lb/hr-sq ft	Polymer	$FeCl_3$ + CaO
1. Primary (P)					
(a) Raw	0.2-0.5	3-5 + 8-12	5.0-8.0	24-26	26-28
(b) Anaerobically digested	0.3-0.6	5-10 + 20-30	2.5-3.0	16-22	18-24
(c) Aerobically digested	0.3-0.6	5-8 + 10-15	2.5-3.0	16-20	18-20
(d) Heat treated	0	0	10.0-12.0	35-45	
2. Waste Activated Sludge (WAS)					
(a) Anaerobically digested	(c)	10-15 + 30-40	1.0-1.5	-	14-18
(b) Aerobically digested	(c)	8-12 + 15-25	1.5-2.0	-	14-18
3. 65% P + 35% WAS (by weight)					
(a) Raw	0.3-0.8	5-8 + 10-15	3.0-3.5	16-18	20-22
(b) Anaerobically digested	(c)	8-12 + 25-35	2.0-2.5	-	15-18
(c) Aerobically digested	(c)	6-10 + 15-20	2.0-2.5	-	16-18
(d) Heat treated	0	0	4.0-6.0	30-35	

(a) Excluding conditioning chemicals
(b) Cationic polymer at 1.5 $/lb
(c) Not recommended
(d) Solids recovery at 95-99%

Table IV lists the expected performance of solid bowl centrifuges for a variety of sludges produced from biological nutrient removal systems. Sizing centrifuges should be based on pilot test results if possible. In the absence of pilot tests, centrifuge manufacturers, who have performance data bases, should be consulted.

(4) Hi-Solids Filter. The Hi-Solids Filter is a relatively recent development that combines vacuum filtration with high pressure cake expression. Filter cake produced from a conventional vacuum filter is further dewatered within a diaphragm press. Operation is continuous, a significant advantage over other high performance devices such as the pressure filter. The diaphragm press may be retrofitted to existing vacuum filters, thereby producing a drier cake that may be incinerated without the need for auxiliary fuel.

Table V lists the expected performance of Hi-Solids filtration for various sludges produced from biological nutrient removal systems. Incinerator ash has also been added to sludge in conjunction with polymer conditioning to enhance the dewatering properties of the sludge. Ash addition, in doses of 50-100% by weight, has resulted in autogenous cake and reduced chemical conditioning cost.

(5) Belt press. The belt press is the most recent mechanical sludge dewatering device to be employed in this country. The belt press utilizes mechanical and shear forces exerted on the sludge cake as it is transported between two endless belts around several rollers to gradually remove moisture from the cake. The advantages of belt presses are very low operating costs, flexible operation, and efficient operation on difficult to dewater sludges.

Table VI lists the expected performance of belt presses for the various sludges produced from biological nutrient removal systems. As the information in Table VI implies, the belt press is capable of a wide range of performance in terms of cake solids. Solids dewatering capacity may be readily traded off with cake solids contents to meet specific needs.

Table IV. Expected Centrifugal Dewatering Performance

Sludge Type	Polymer - wt. % (a)	Cake Solids - wt. % TS
1. Primary (P)		
(a) Raw	0.1-0.2	25-30
(b) Anaerobically digested	0.2-0.4	20-25
(c) Aerobically digested	0.2-0.3	18-22
(d) Heat treated	0	30-35
2. Waste Activated Sludge (WAS)		
(a) Anaerobically digested	0.3-0.5	12-14
(b) Aerobically digested	0.2-0.4	10-12
3. 65% P + 35% WAS (by weight)		
(a) Raw	0.2-0.3	15-20
(b) Anaerobically digested	0.3-0.4	15-20
(c) Aerobically digested	0.2-0.4	14-18
(d) Heat treated	0-0.1	25-30

(a) Cationic polymer at 1.5 $/lb
(b) Solids recovery at 90-95%

Table V. Expected Hi-Solids Filter Performance

Sludge Type	Chemical Conditioning, wt. % $FeCl_3$ + CaO	Filter Yield(a) lb/hr-sq ft	Cake Solids wt. % TS
1. Primary (P)			
(a) Raw	3-5 + 8-12	4.0-5.0	35-40
(b) Anaerobically digested	5-10 + 20-30	2.5-3.0	30-35
(c) Aerobically digested	5-8 + 10-15	2.5-3.0	25-30
(d) Heat treated	0	5.0-8.0	50-60
2. Waste Activated Sludge (WAS)			
(a) Anearobically digested	10-15 + 30-40	1.0-1.5	18-23
(b) Aerobically digested	8-12 + 15-25	1.5-2.0	20-25
3. 65 % P + 35% WAS (by weight)			
(a) Raw	5-8 + 10-15	2.5-3.0	27-32
(b) Anaerobically digested	8-12 + 25-35	2.0-2.5	27-32
(c) Aerobically digested	6-10 + 15-20	2.0-2.5	24-28
(d) Heat treated	0	4.0-6.0	50-55

(a) Excluding conditioning chemicals
(b) Solids recovery at 95-98%

Table VI. Expected Belt Filter Performance

Sludge Type	Polymer - wt. % [a]	Cake Solids - wt. % TS
1. Primary (P)		
(a) Raw	0.1-0.2	25-40
(b) Anaerobically digested	0.1-0.3	20-30
(c) Aerobically digested	0.2-0.3	20-28
(d) Heat treated	0	40-50
2. Waste Activated Sludge (WAS)		
(a) Anaerobically digested	0.3-0.5	18-22
(b) Aerobically digested	0.3-0.5	18-25
3. 65% P + 35% WAS		
(a) Raw	0.2-0.3	22-30
(b) Anaerobically digested	0.2-0.4	18-25
(c) Aerobically digested	0.2-0.4	20-28
(d) Heat treated	0	40-45

(a) Cationic polymer at 1.5 S/1b
(b) Solids recovery at 90-98%

The belt press could well be considered a major breakthrough in biological sludge processing technology. It has the capability to readily and efficiently dewater waste activated sludge which has heretofore proved to be a difficult and costly application for other mechanical devices. While centrifuges and vacuum filters have been applied to the waste activated sludge dewatering application, the operation is generally unattractive in terms of low cake solids content, difficulty of operation and high operating costs. The experience to date indicates that belt presses overcome these problems.

As with centrifuges, sizing belt presses should be based on pilot tests when practical.

(6) Pressure filtration. Filter pressing is a batch operation where sludge is pumped into chambers for dewatering at pressures of 100-250 psig. Pressure filtration is capable of higher cake solids concentrations than vacuum filtration and centrifugation. Pressure filtration has seen limited use, however, since its operation is somewhat complex and costly. A further disadvantage of pressure filtration is that, being a batch operation, it requires cake breaking and storage facilities, particularly if cake incineration is practiced.

Table VII lists the expected performance of pressure filtration for various sludges. Conditioning is normally accomplished with the addition of inorganic chemicals. Incinerator ash has also been added to sludge to enhance its dewatering properties. Ash addition, in doses of 100-200% by weight, has resulted in drier cakes, reduced chemical conditioning requirements and improved cake discharge.

Sizing filter presses should be based on pilot tests, when practical.

Reduction

The primary purpose of sludge reduction is to reduce the quantity of waste sludge to be disposed of on land. Sludge reduction has been employed in situations where land disposal sites have been limited or costly, and where engineering judgement and/or regulatory restrictions concering land disposal operations have indicated the desirability of sludge reduction.

Table VII. Expected Filter Press Performance

Sludge Type	Chemical Conditioning, wt. % $FeCl_3$ + CaO	Cake Solids - wt. % TS
1. Primary (P)		
(a) Raw	3-5 + 8-12	40-45
(b) Anaerobically digested	5-10 + 20-30	33-38
(c) Aerobically digested	5-8 + 10-15	25-30
(d) Heat treated	0	45-50
2. Waste Activated Sludge (WAS)		
(a) Anaerobically digested	10-15 + 30-40	22-27
(b) Aerobically digested	8-12 + 15-25	20-25
3. 65% P + 35% WAS		
(a) Raw	5-8 + 10-15	27-32
(b) Anaerobically digested	8-12 + 25-35	30-35
(c) Aerobically digested	6-10 + 15-20	23-28
(d) Heat treated	0	40-45

(a) Solids recovery at 95-99%

Thermal reduction, i.e., incineration, is the principal sludge reduction method. Incineration results in evaporation of the moisture content of the sludge and combustion of the volatile fraction of the sludge solids. Other sludge processing unit operations which may be considered as sludge reduction processes are digestion and heat treatment. However, sludge stabilization and/or conditioning are the principal purposes of these processes and they will not be discussed here.

Through the 60's and early 70's, incineration practice as a sludge reduction technique increased substantially. This was especially true for multiple hearth incineration because of its simplicity, flexibility and effectiveness. Increasing land costs, decreasing land availability and more stringent standards for land disposal were responsible for the trend toward incineration.

The principal disadvantage of sludge incineration is the cost for auxiliary fuel which must be added to the incinerator for sludge combustion. The heat generated from sludge combustibles during incineration is usually insufficient to sustain the incineration process if the solids content of the cake is less than about 30%. Supplemental heat, in the form of natural gas or fuel oil, must be added. With the energy crisis, reduced availability and rapid rise in cost of fuels, more sophisticated dewatering techniques capable of dewatering sludge to solids contents near or beyond the autogenous point have been empahsized. Filter presses, Hi-Solids filters and belt presses are the dewatering methods which have received more attention in this regard. Figure 2 presents auxiliary fuel requirements for various cake conditions and illustrates the importance of designing dewatering systems to achieve near maximum cake solids content.

SLUDGE PROCESSING AND DISPOSAL ALTERNATIVES

Biological nutrient removal systems were examined relative to the sludges produced from those systems and the methodology associated with ultimate disposal of the sludges. Biological nutrient removal systems were classified as those appropriate for nitrification, nitrogen removal and nitrogen and phosphorus removal. A number of alternative sludge processing schemes were developed for each biological nutrient removal system classification and

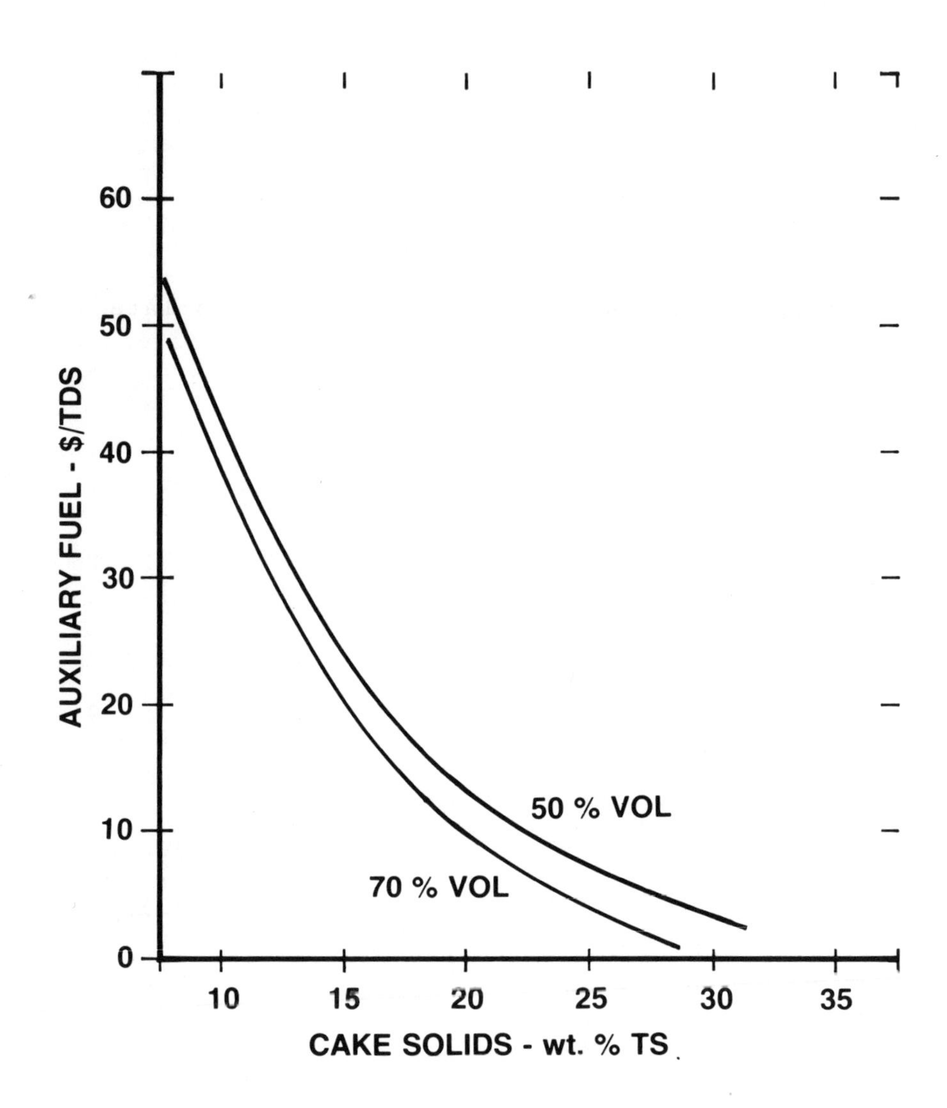

FIGURE 2. INCINERATION FUEL REQUIREMENTS

evaluated in terms of their appropriateness.

Nitrification

Biological nitrification may be accomplished with single stage extended aeration systems such as Carrousel, or two stage systems with the first stage designed for carbonaceous removal and the second stage for nitrification. Figure 3 schematically describes these treatment options. These options, including whether or not primary clarifiers are employed impact on the selection of an appropriate sludge processing flow sheet.

Two basic situations exist relative to the final disposition of sludge. The first involves an availability of land application or disposal sites for sludge disposal. The second involves an unavailable or distant or limited capacity of land application or disposal sites. In the latter situation, sludge incineration is the only practical means of reducing sludge quantities to minimize the cost of sludge disposal.

Figure 4 describes the sludge processing scheme for the biological nitrification flow sheet when sludge reduction by incineration is necessary.

The strategy is to provide a system which maximizes the solids content of the cake going to incineration so as to maximize the need for auxiliary fuel. Utilizing devices capable of dewatering sludge at or near its autogenous point is essential. Pressure filters, Hi-Solids filters and belt presses are the logical alternatives. The cost effectiveness of these alternatives should be evaluated with respect to the specific conditions which exist at each installation.

Regardless of which sludge dewatering alternative is selected, the strategy of maximizing cake solids content is furthered by minimizing the quantity of biological sludge and maximizing the quantity of primary sludge to be dewatered. In this regard, it would be desirable to utilize primary clarification in the biological nitrification system flow sheets described on Figure 3. Stabilization by digestion of either primary or secondary sludge should not be practiced since a reduction in the sludge volatile content, and, thereby, the BTU value of the sludge would occur.

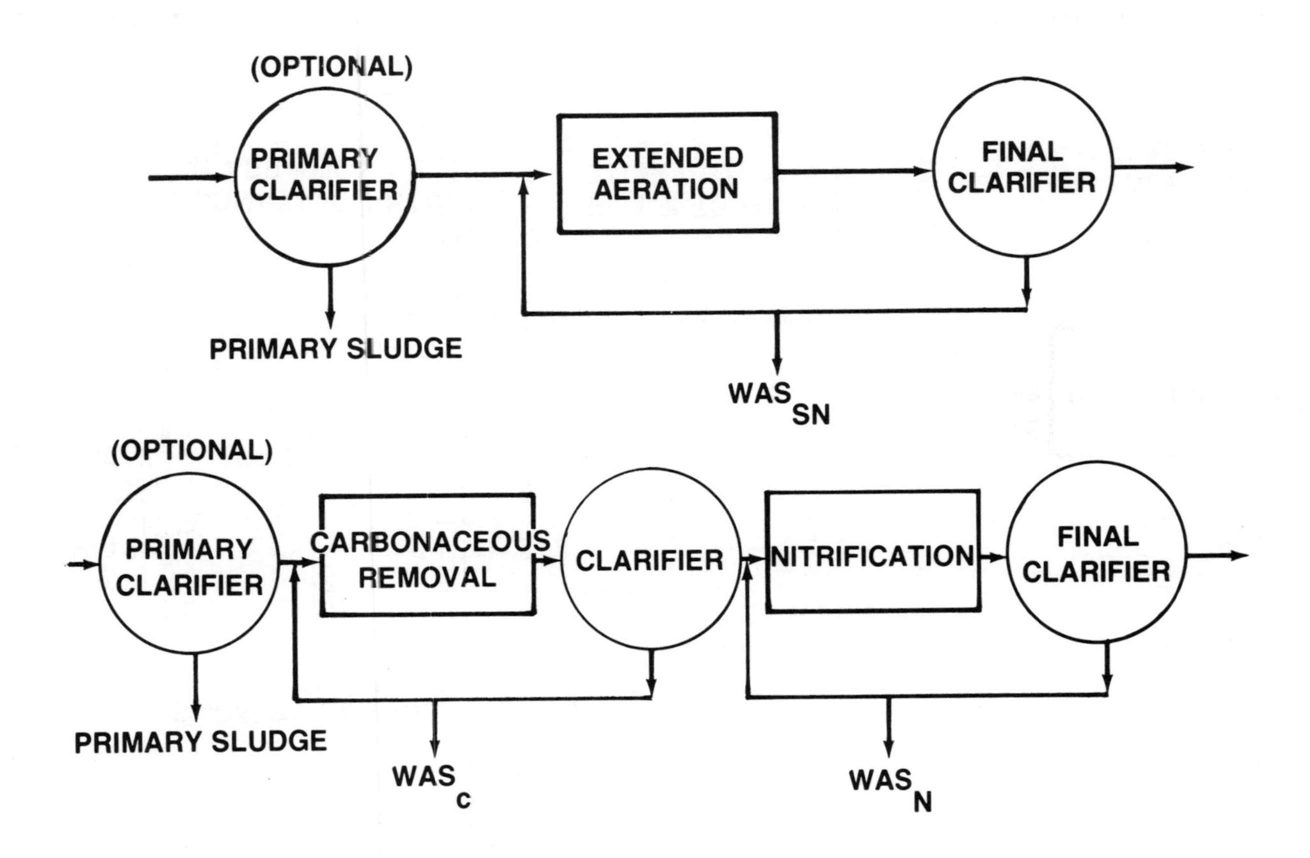

FIGURE 3. BIOLOGICAL NITRIFICATION SYSTEM OPTIONS

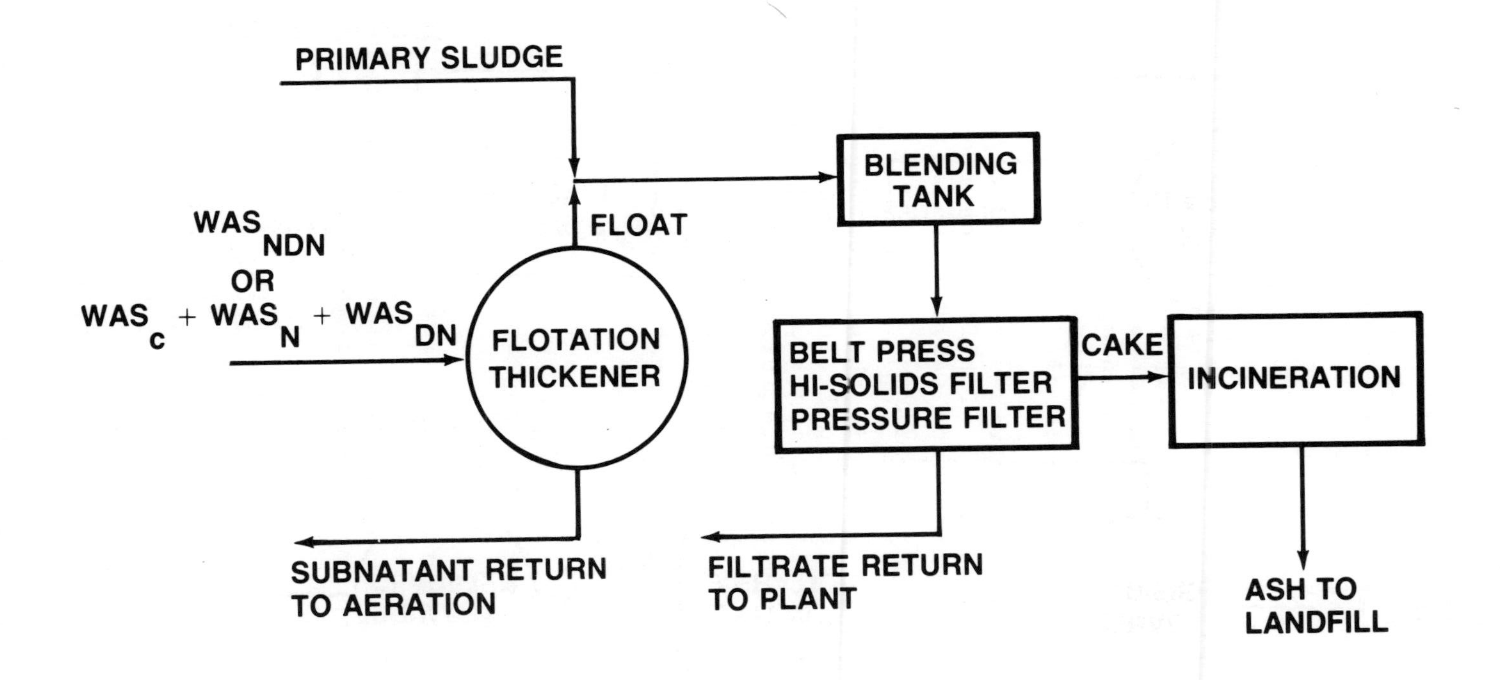

FIGURE 4. SLUDGE PROCESSING SCHEME WITH INCINERATION FOR BIOLOGICAL NITRIFICATION FLOW SHEETS

Thickening of the biological sludges is necessary to enhance the performance of the dewatering equipment. Increased capacity and cake solids content and reduced chemical conditioning costs generally result from thickening feed sludges. The cost effectiveness of flotation thickening versus gravity thickening is anticipated for the system shown on Figure 4 since the necessity of incineration implies a large facility.

A sludge blending tank should be provided in the system shown on Figure 4 to minimize fluctuations in dewatering performance due to feed sludge variations. Full advantage of the efficacy of primary sludge on dewatering performance would be insured with a blended sludge.

Figure 5 describes the sludge processing scheme for the biological nitrification flow sheet for the situation where land application or disposal sites are readily available.

The strategy in this situation is to process the waste sludges only to a degree consistent with the physical handling characteristics and/or solids content requirements for disposal.

Ultimate disposal methods requiring a dewatered sludge (cake) would necessitate the inclusion of one of the sludge dewatering alternatives listed on Figure 5 into the solids processing flow sheet. The cost effectiveness of these alternatives should be evaluated with respect to the specific conditions which exist at each installation. Sludge haul distance, and engineering and regulatory restraints are among the factors which would be considered.

Ultimate disposal methods not requiring a cake could eliminate the need for dewatering and would permit the application of thickened, stabilized sludge to land. Consideration of economics could, however, indicate the desirability of dewatering.

Since most state and federal regulations require that sludge be stabilized prior to land disposal, it would be economically desirable to eliminate primary clarification systems shown on Figure 3. Within the context of the sludge processing flow sheet, the single stage nitrification system has advantages over the two stage system. The waste

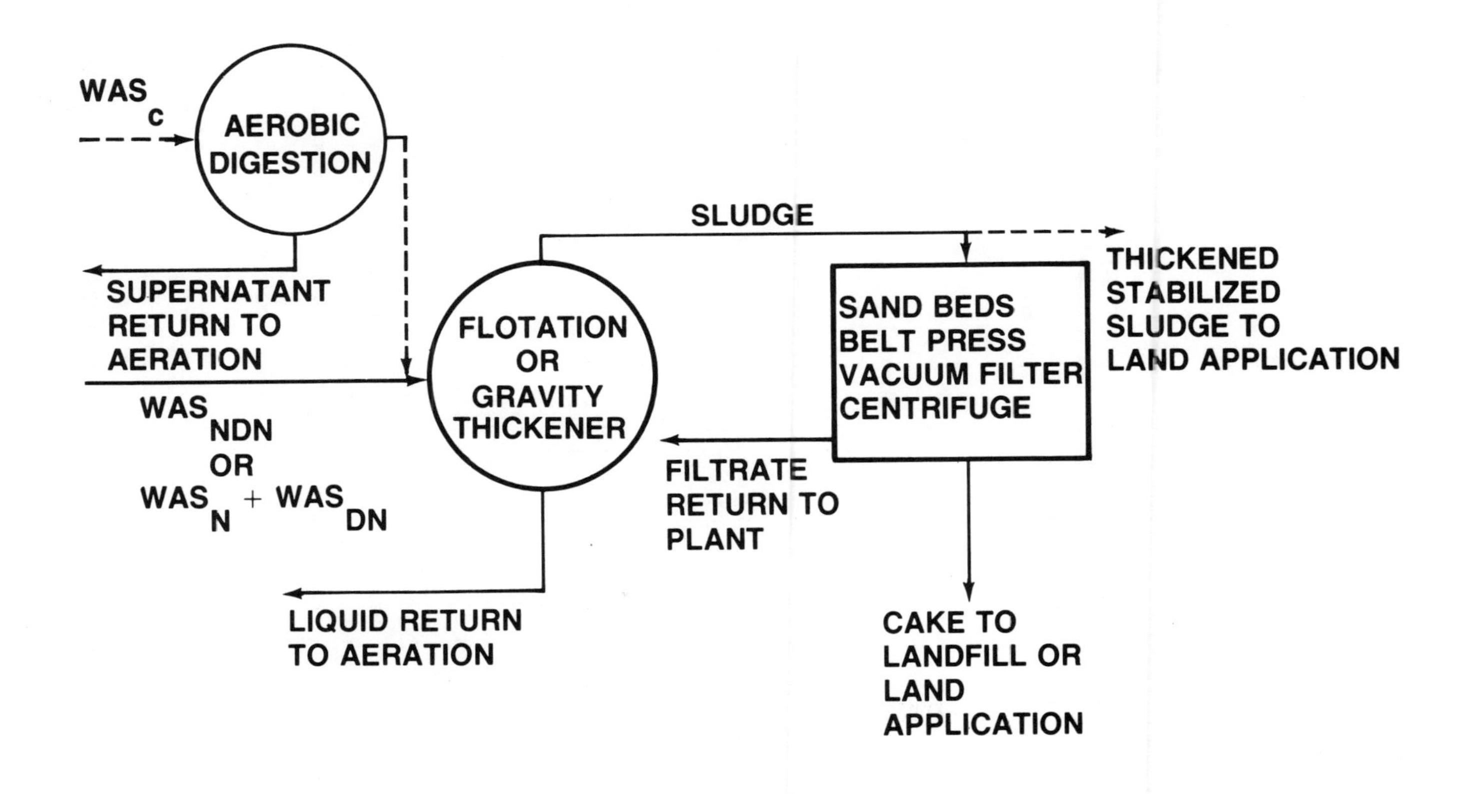

FIGURE 5. SLUDGE PROCESSING SCHEME WITHOUT INCINERATION FOR BIOLOGICAL NITRIFICATION FLOW SHEETS

biological sludge produced from the single stage system is considered to be stabilized, whereas the waste biological sludge produced from the carbonaceous removal stage of the two stage system requires separate stabilization. In this regard, the single stage biological nitrification flow sheet, without primary clarifiers, results in the simplest and least costly sludge processing scheme.

Flotation and gravity thickening of the waste biological sludge should be evaluated in light of the specific circumstances existing at each installation. Generally, flotation thickening can not be economically justified, in spite of its superior performance capability, for plants smaller than 1-2 MGD.

Nitrogen removal

Biological nitrogen removal may be accomplished using either the Bardenpho Process, Carrousel Process or a three stage process as schematically shown on Figure 6.

Figure 7 describes the sludge processing scheme for the biological nitrogen removal flow sheet when sludge reduction by incineration is required.

For reasons discussed earlier, the strategy is to provide devices capable of dewatering sludge at or near its autogenous point. As before, pressure filters, Hi-Solids filters and belt presses are the appropriate alternatives. It would be advantageous to utilize primary clarification in the biological nitrogen removal flow sheets described on Figure 6 in order to enhance the dewatering characteristics of the sludge produced. For reasons discussed earlier, stabilization of either the primary or biological sludge would not be practiced.

Thickening of the mixed biological sludges prior to dewatering would be required. The cost effectiveness of flotation thickening is anticipated for the system shown on Figure 7 since the necessity of incineration implies a large facility.

Figure 8 describes the sludge processing scheme for the biological nitrogen removal flow sheet for the situation where land application or disposal sites are readily available.

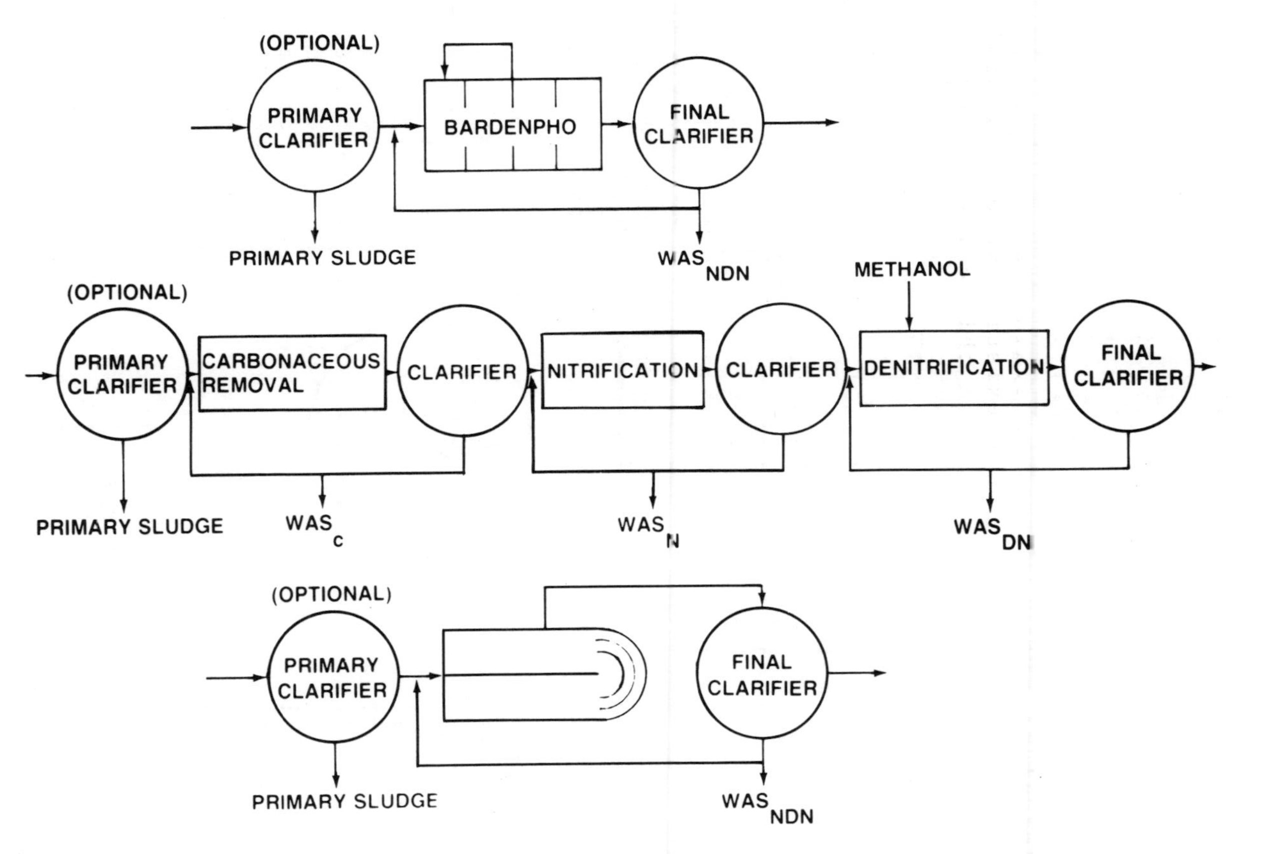

FIGURE 6. BIOLOGICAL NITROGEN REMOVAL SYSTEM OPTIONS

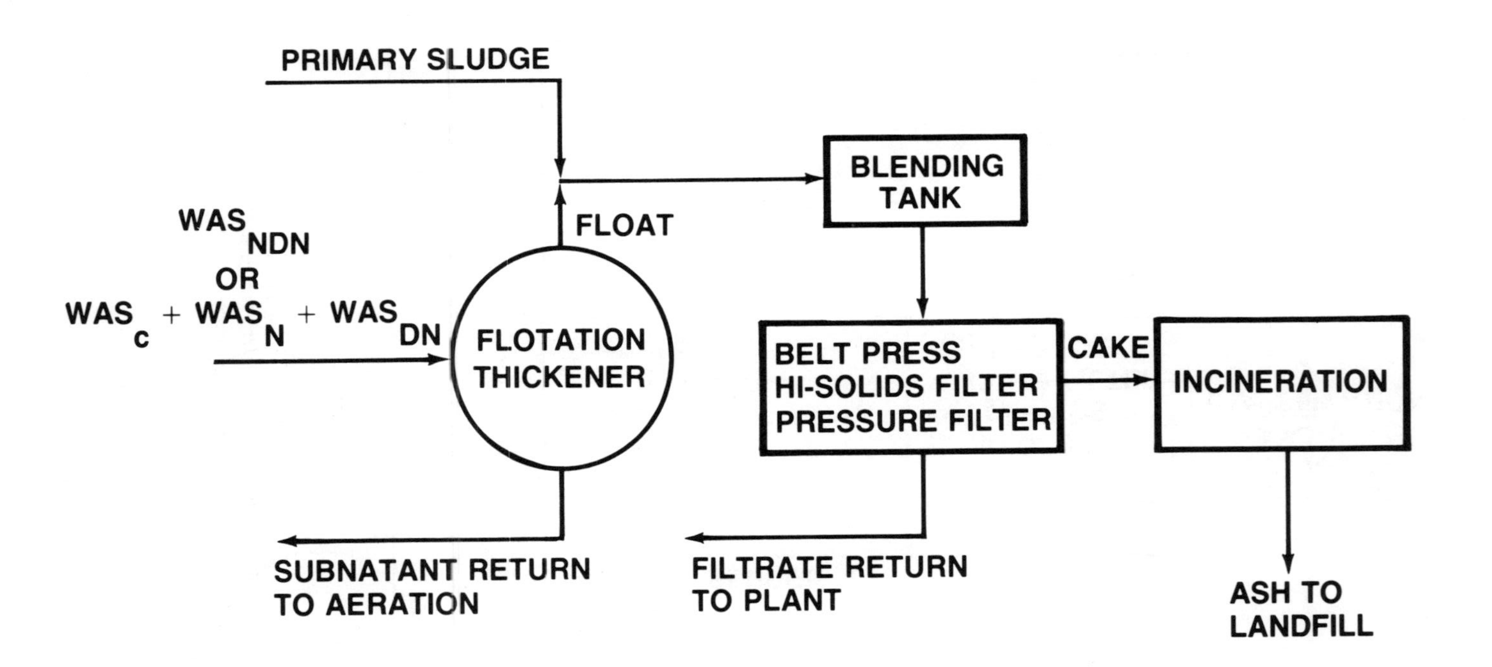

FIGURE 7. SLUDGE PROCESSING SCHEME WITH INCINERATION FOR BIOLOGICAL NITROGEN REMOVAL FLOW SHEETS

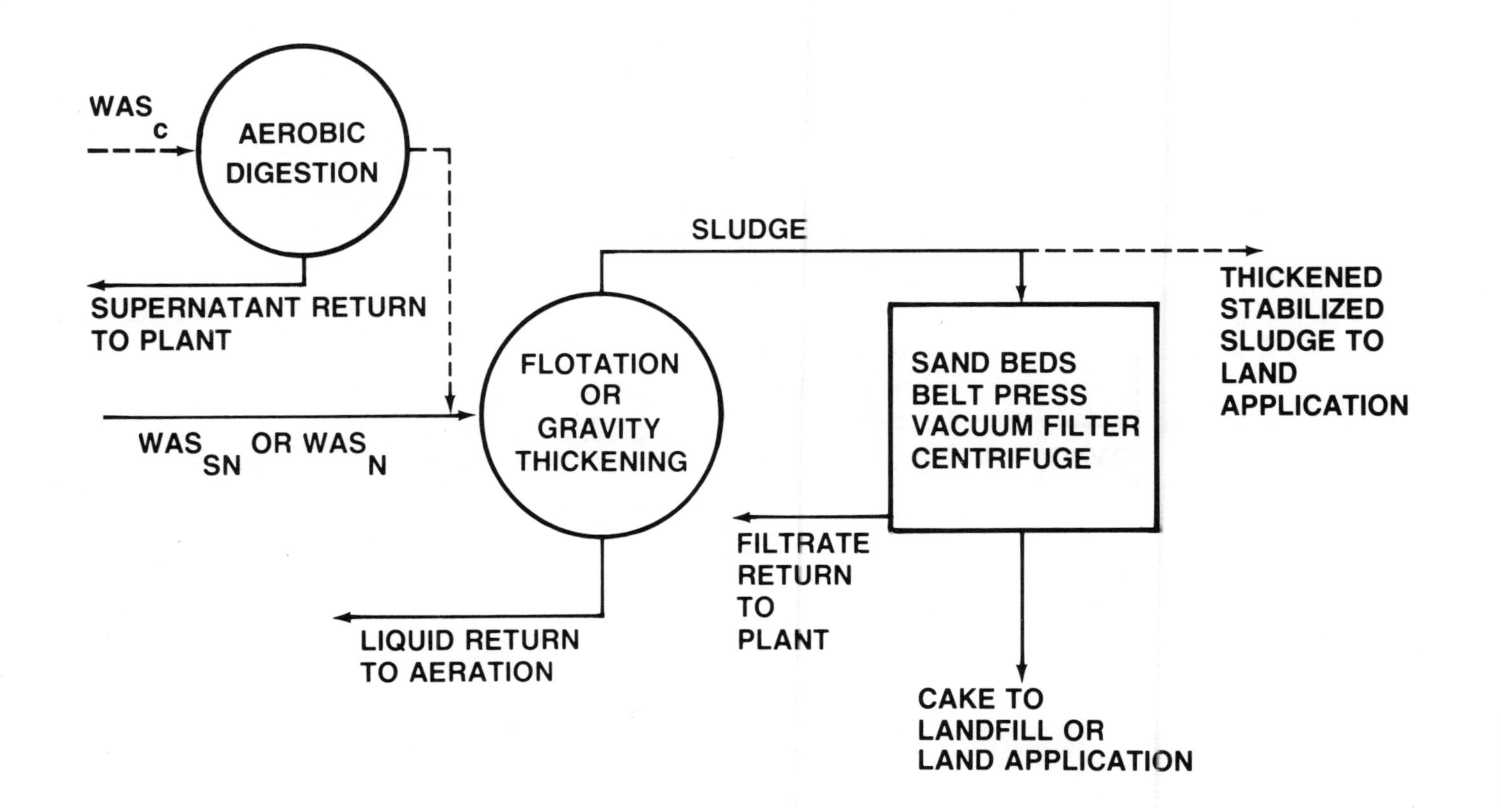

FIGURE 8. SLUDGE PROCESSING SCHEME WITHOUT INCINERATION FOR BIOLOGICAL NITROGEN REMOVAL FLOW SHEETS

For reasons discussed earlier, the strategy in this situation would be to process the waste sludges only to a degree consistent with the physical handling characteristics and/or solids content requirements for disposal. For situations where dewatering is advantageous, the dewatering alternatives listed on Figure 8 would be evaluated with respect to specific conditions for each application. For situations where dewatering is not necessary, thickened, stabilized sludge would be disposed of on land.

Flotation and gravity thickening of the waste biological sludge should be evaluated for treatment and cost effectiveness. Although flotation has a performance capability superior to gravity thickening, flotation thickeners are seldom employed in plants smaller than 1-2 MGD.

Within the context of the sludge processing flow sheet, it would be advantageous to eliminate primary clarification from the biological nitrogen removal systems shown on Figure 6. Doing so would eliminate the need for stabilizing primary sludge prior to land disposal. As implied on Figure 8, the Bardenpho and Carrousel systems are advantageous since stabilized sludges are produced from these systems. The three stage process shown on Figure 6 results in the production of biological sludge from the carbonaceous removal stage which requires separate stabilization. In this regard, the Carrousel or Bardenpho biological nitrogen removal systems, without primary clarifiers, result in the simplest and least costly sludge processing scheme.

Nitrogen and phosphorus removal

Biological nitrogen and phosphorus removal may be accomplished using staged treatment systems such as the Bardenpho Process as schematically shown on Figure 9.

Figure 10 describes the sludge processing scheme for the biological nitrogen and phosphorus removal flow sheet for the situation where sludge reduction by incineration is required. The sludge processing flow sheet is identical to that shown for the biological nitrogen removal application on Figure 7.

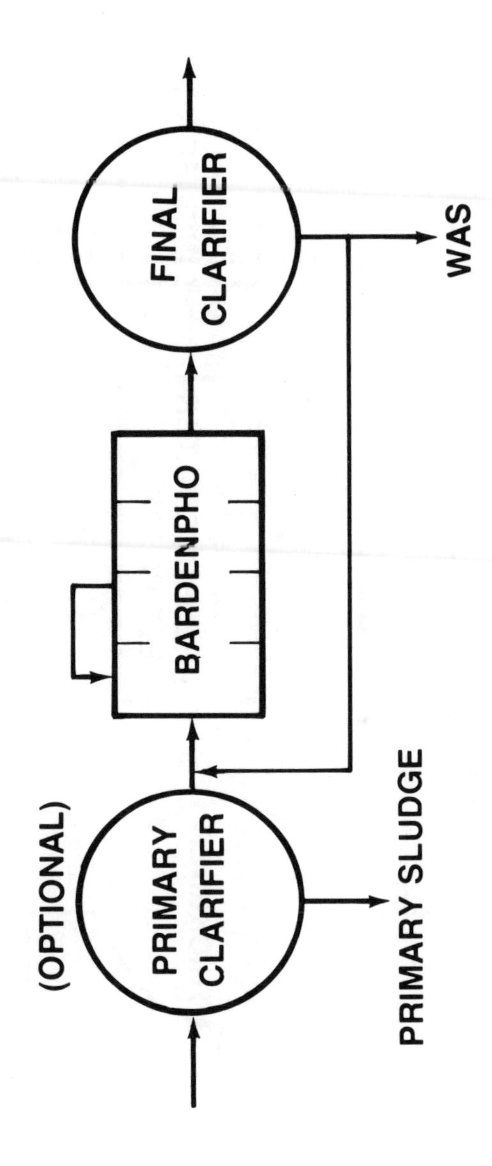

FIGURE 9. BIOLOGICAL NITROGEN AND PHOSPHORUS REMOVAL SYSTEM

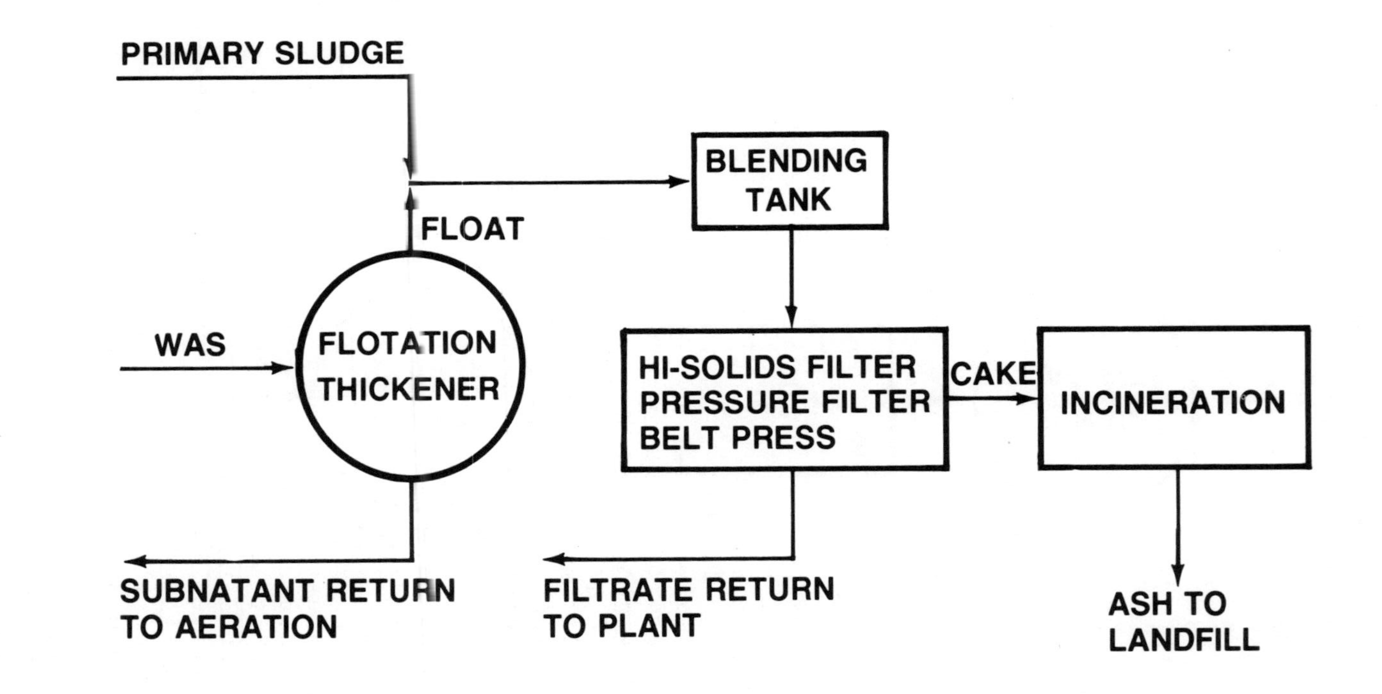

FIGURE 10. SLUDGE PROCESSING SCHEME WITH INCINERATION FOR BIOLOGICAL NITROGEN AND PHOSPHORUS REMOVAL FLOW SHEETS

In the context of the biological nitrogen and phosphorus removal system, flotation thickening would be the method of choice. Utilizing flotation thickening would insure that phosphorus would not be stripped from the biological sludge due to anaerobic conditions prior to being dewatered.

Figure 11 describes the sludge processing scheme for the biological nitrogen and phosphorus removal system for the situation where land application or disposal sites are readily available. The sludge processing flow sheet is identical to that shown for the biological nitrogen removal application on Figure 8. Flotation thickening would be the method of choice to eliminate the possibility of phosphorus stripping due to anaerobic conditions which could exist within gravity thickeners.

SUMMARY AND CONCLUSIONS

Methods of processing waste sludges produced from biological nutrient removal systems were examined. Three categories of biological nurrient removal systems considered were those capable of nitrification, nitrogen removal, and nitrogen and phosphorus removal. Sludge processing schemes were developed for each category in light of the characteristics of the sludges produced and the method of ultimate disposal of the sludge. Sludge processing schemes were further categorized in terms of the need for sludge reduction by incineration prior to land disposal.

For those situations where incineration is required, a sludge processing scheme such as that shown on Figure 10 was determined to be most appropriate for either biological nutrient removal category. The pressure, filter, Hi-Solids filter and belt press, devices capable of dewatering sludge to the extent that auxiliary fuel requirements are minimized, are suggested as the pertinent dewatering alternatives requiring economic evaluation for each design situation. In order to enhance the performance of the dewatering equipment and to minimize the cost of the dewatering incineration system, it is recommended that primary clarification be employed in the main plant flow sheet. This would reduce the quantity of biological sludge and result in the production of primary sludge

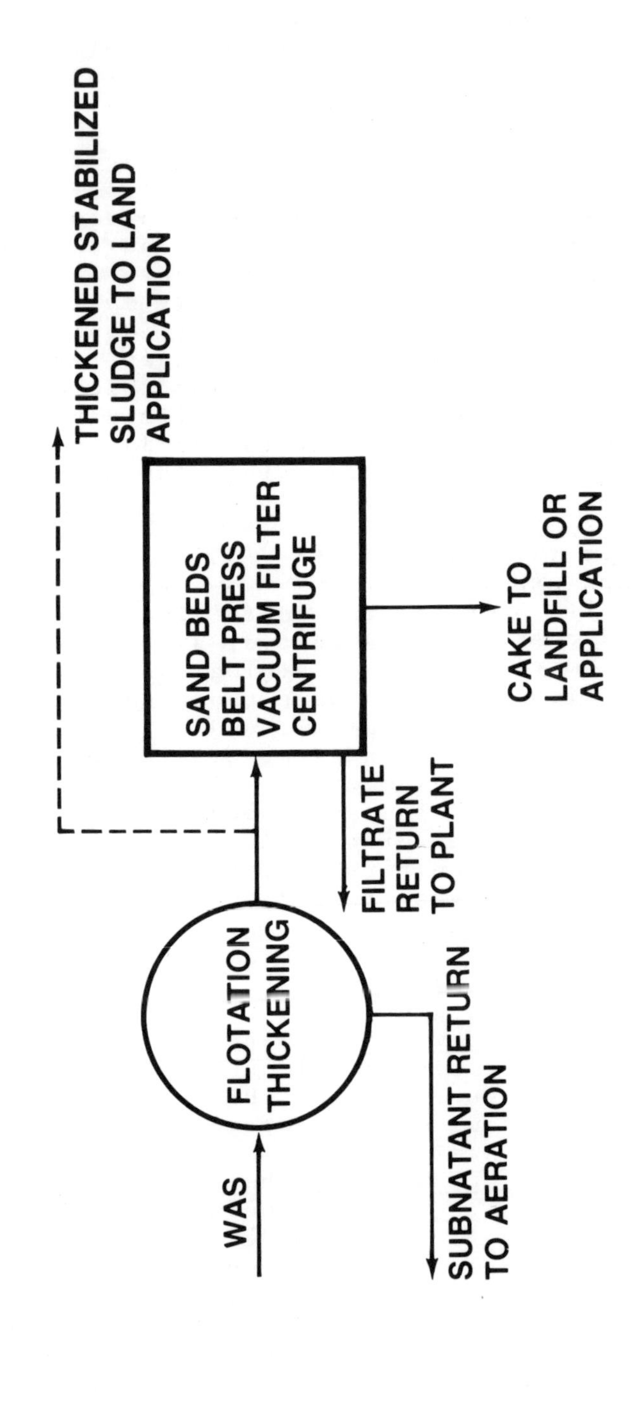

FIGURE 11. SLUDGE PROCESSING SCHEME WITHOUT INCINERATION FOR BIOLOGICAL NITROGEN AND PHOSPHORUS REMOVAL FLOW SHEETS

which has dewatering characteristics superior to biological sludge. Waste biological sludge would be flotation thickened and blended with the primary sludge prior to dewatering.

For those situations where sludge reduction by incineration is not required prior to land disposal, a sludge processing scheme such as that shown on Figure 8 was determined to be most appropriate. Primary clarification is not recommended for the main plant flow sheet since the primary sludge produced would have to be stabilized (i.e., digestion) prior to land disposal. Similarly, biological nutrient removal systems producing stabilized sludges are advantageous since the need for separate stabilization is eliminated. The stabilized, biological sludges would be thickened by gravity or flotation depending upon economics and local conditions. After thickening, the sludges would be disposed of directly to the land or depending upon economics, regulations and local conditions, dewatered prior to disposal. Sand beds, the belt press, vacuum filter and centrifuge are the pertinent dewatering devices requiring economic evaluation for each design situation.

12. COST EFFECTIVE WASTEWATER TREATMENT

John D. Wright. Partner, Hazelet & Erdal Consulting Engineers, Louisville, Kentucky

ABSTRACT

The 20 year water pollution control plan for the City of Campbellsville, KY, requires construction of a 4.2 MGD wastewater treatment plant containing a BOD averaging 350 mgl, or a total of 12,245 lbs. per day. Effluent requirements to the stream are 10 mgl BOD and 20 mgl suspended solids. As a result of EPA's recent concern for cost effectiveness, the plant was designed with value analysis applied to each treatment function.

The low bid for the project received on November 22, 1977 totaled $3,520,800. This is equal to a unit cost of $287 per lb. of BOD or $47.92 per population equivalent.

INTRODUCTION

Recently EPA introduced value engineering (value analysis) into the management of their federal grant programs covering design and construction of wastewater treatment facilities. Early on, many consultants responded with the general expression such as "We have always been trying to be cost effective in our designs". In commencing with the design of wastewater treatment facilities for the City of Campbellsville we decided at the outset that we would use the value analysis approach throughout the design phase, hopeful that we could control treatment plant construction costs. At that time it was generally understood that secondary treatment was costing in the magnitude of $1.00 to $1.50 per gal. of wastewater, roughly equal to $100 to $150 per population equivalent. It is the purpose of this paper to briefly summarize our cost effective approach and the results thereof. Early in the design phase it was realized that we should divide the total treatment process into basic treatment functions, such as pumping raw wastewater, activated sludge treatment, returning activated sludge, chlorinating final effluent, and etc. In other words, we developed the thesis that functional processes within the total treatment plant process should each be made cost effective.

VALUE ANALYSIS

Team #7 of the Value Analysis Workshop[1] conducted by ACEC and AIA in Boston during February 24 - March 1, 1974 reduced the cost of a 4.5 MGD wastewater treatment plant from $4,155,000 to $2,476,000, a savings of $1,679,000 or 40 percent. The typical approach to proceeding with the analysis of any material, procedure or technique is to question the function of the material, procedure, or technique by asking: What is it? - What does it do? - What does it cost? - What is it worth? - What other material, procedure, or technique could do the same job? Early in its course, Team #7 asked "Have we the authority to change the basic wastewater treatment plant process?" (The word "process" connoting "procedure or "technique"). With permission granted, Team #7 was able to reduce the treatment plant cost by 40 percent, principally by process change.

In addition to first cost analysis of alternate methods of performing a function, value analysis must include parallel investigation into the operation and maintenance costs of alternate methods. For example, it is fundamental that higher first cost in building insulation yields a lower life cycle cost because of lower heating bills.

In determining alternate methods of performing a given function, the value analyst must endeavor to have a completely open and inquiring mind. He must determine if a function is actually necessary. We are reminded that habits or customs work more constantly and with greater force than reason. Attitudes can rob us of value because they suggest the continuation of existing habits. Value analysts have often determined lack of value in a given design because of honest wrong beliefs by the designer.

From the beginning we addressed ourselves to the arrangement of facilities to permit party wall construction as a matter of economy. Piping arrangements were given scrutiny to obtain the most optimum layouts. Valves, pumps, and meters were evaluated to be certain proper function was obtained, but with lowest first cost. Also, we analyzed the plant flow diagram to determine the number of basic treatment plant functions actually required. In considering these many factors, we viewed the developing plant layout from the point of view of the future plant operator, and endeavored to develop a facility that we ourselves could and would enjoy operating. Finally, the basic activated sludge process was selected only after combined consideration of first cost and future energy requirements. A visit to several Carrousel treatment plants in Europe led us to the conclusion that the Carrousel process would be the most cost effective and energy effective activated sludge process for Campbellsville, KY.

TREATMENT PLANT FUNCTIONS

The Campbellsville wastewater treatment plant was required to treat 4.2 MGD of municipal and industrial wastewater containing a design total of 12,445 lbs. of BOD equal to an average strength of 350 mgl. Effluent requirements are 10 mgl BOD and 20 mglSS equal to 98 percent removal of the pollutants. The basic plant functions were: (1) Solids Comminution, (2) Grit Removal, (3) Raw wastewater pumping, (4) Carrousel Activated Sludge, (5) Final Clarification, (6) Activated Sludge Return, (7) Skimmings Removal, (8) Effluent Chlorination, and (9) Waste Solids Disposal.

PLANT FLOW DIAGRAM

Figure 1 shows the basic layout and flow diagram. The 21 inch and 27 inch interceptors discharge into the raw wastewater pumping station which contains comminutors, raw wastewater pumps and grit pumps. The pumped discharge may enter each of the three Carrousel tanks. At the right end of the Carrousel tanks, effluent weirs discharge to a distribution box which splits the flow to the three clarifiers. Clarifier effluent is piped through the 30 inch line to the chlorine contact chambers, thence discharged to the stream. Return solids are pumped from the clarifiers through the 16 inch line back to the front of the Carrousel tanks. Excess or waste activated solids are discharged from the 16 inch line to a flocculation tank, thence to a thickener tank. Skimmings are pumped from the clarifiers to a manually cleaned skimmings decant tank.

COST-EFFECTIVE DESIGN CONCEPTS

(1) Solids Communition - Figure 2 shows two comminutors, arranged in separate chambers to permit easy isolation for maintenance and repair. If a comminutor fails and surcharge occurs, the surcharge simply rises up and passes through a bar rack. This arrangement eliminates the expenditure of premium time (overtime) or weekend labor to repair the comminutor. The first cost of comminution facilities in the plant totaled \$36,000, equal to \$0.49 per population equivalent (P.E.).

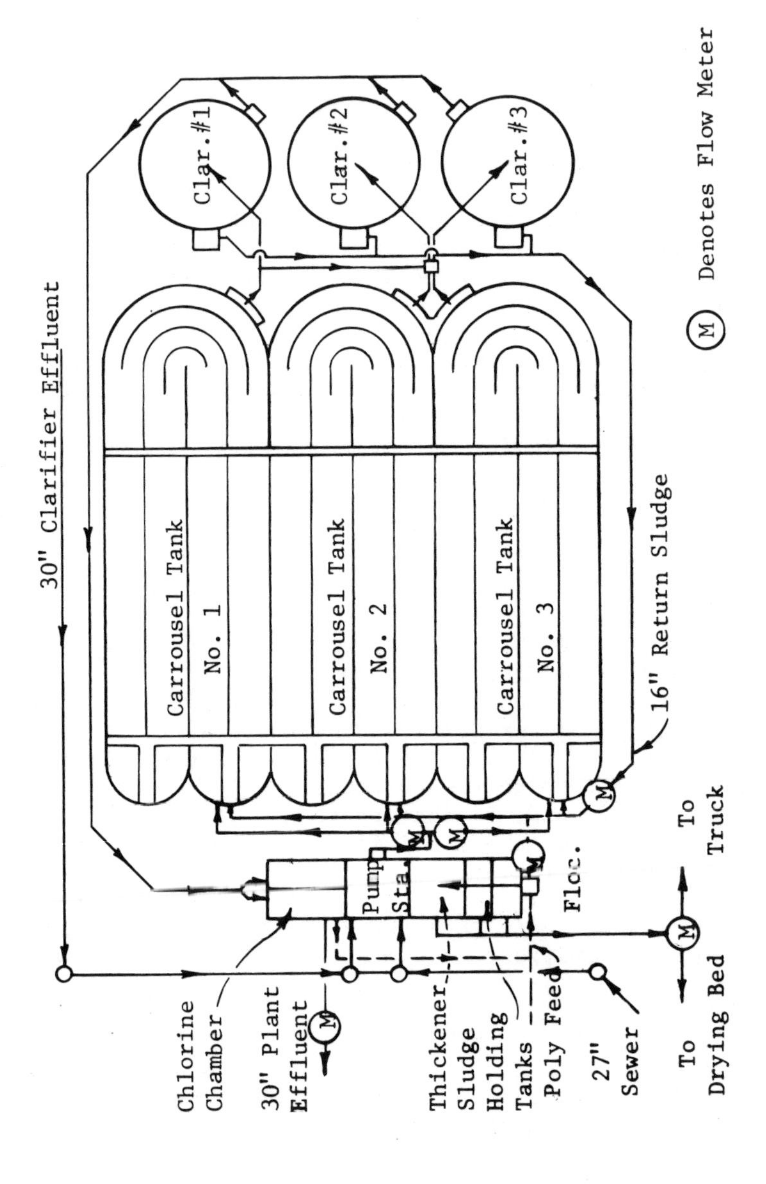

FIGURE 1. Process Flow

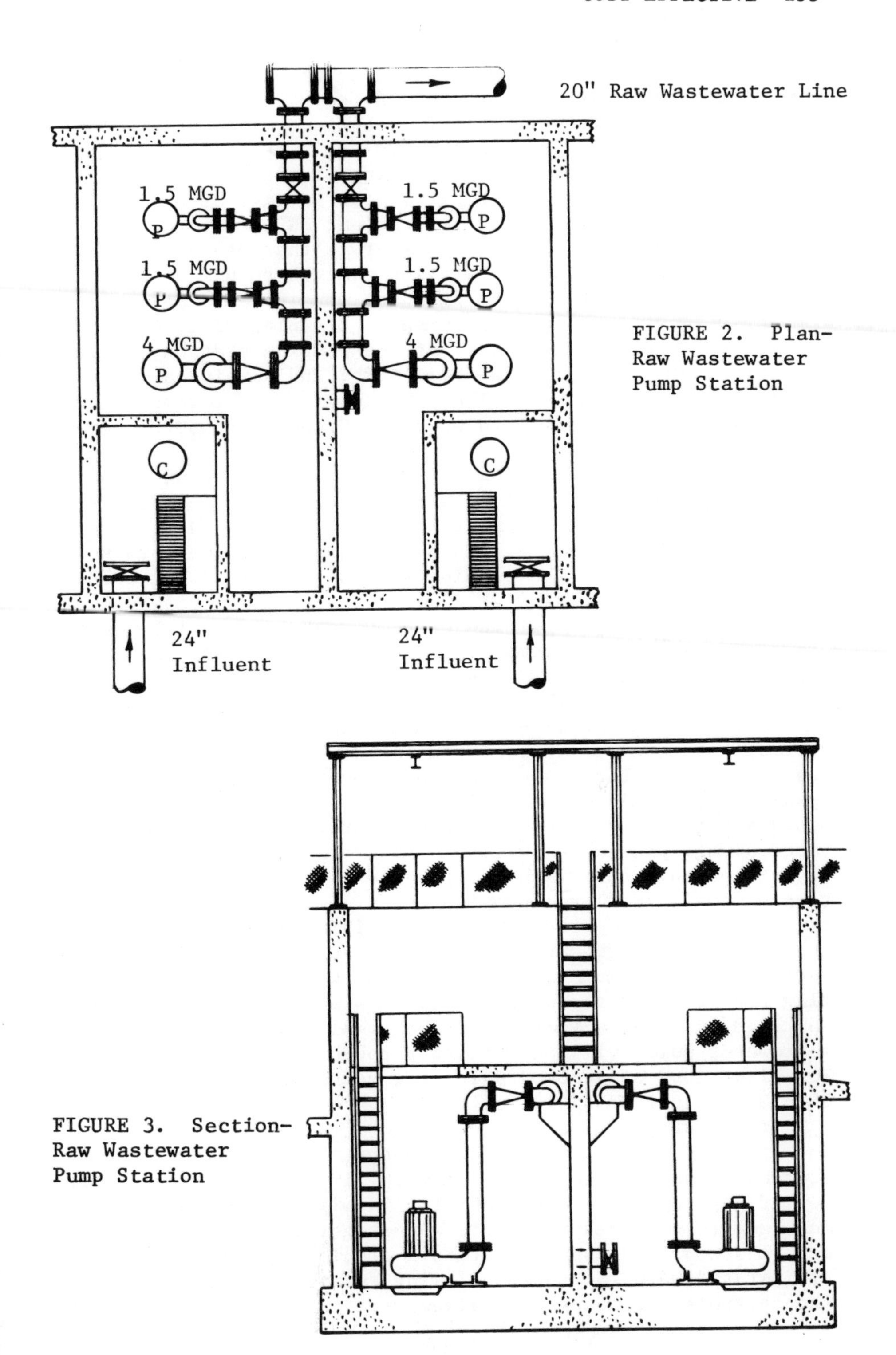

FIGURE 2. Plan-
Raw Wastewater
Pump Station

FIGURE 3. Section-
Raw Wastewater
Pump Station

(2) Grit Removal - The two halves of the pumping station have flat bottoms with an area of 16 feet x 32 feet or 512 square feet. Accordingly, the flat bottoms with the pumping station _is_ the basic grit chamber for this facility. At 3 inches average grit depth, the pumping station would hold approximately 250 cu. feet of grit. In 90 days at design flow of 4.2 MGD at 1/2 cu. feet grit per MGD, grit volume would equal 189 cu. feet. The answer is that manual cleaning of the floor of the pumping station every 3 to 6 months is labor effective, rather than spending considerable money for conventional grit removal equipment. Cleaning will be accomplished by dewatering one half of the pumping station, hosing the pump chamber floor to a small sump, and pumping a grit slurry to the drying bed. The first cost of grit removal facilities in the plant totaled $15,800, equal to $0.22 per P.E.

(3) Raw Wastewater Pumping - The basic function of a pumping station is to lift liquid. It requires a container to receive the liquid, pumps to pump the liquid and pipes or conduits to discharge the liquid to a higher elevation. Refer to Figures 2 and 3. The pumping station consists of two rectangular halves, each 16' x 32' with a total operating depth of 10 feet. Each half of the station contains two 1.5 MGD pumps and one 4.0 MGD pump. All pumps are submersible. Total capacity of the station is 14 MGD, equal to 333 percent of design flow. Any of the 6 submersible pumps can be removed in a matter of a few minutes by motor operated hoist. No consideration was given to variable speed pumping units for two basic reasons. Fixed speed pumps, properly selected for total dynamic head, will be more efficient. Also, vari-drive controls would add to the cost of electrical work. Moreover, the activated sludge process selected is not affected by sudden change in rate of influent discharge produced by intermittent operation of small and large pumping units. After all, the function of the plant is to treat all flow arriving at the plant. If the activated sludge process would be so finicky to require variable drive pumping, then we would consider we were selecting the wrong treatment process. Raw wastewater pumps will be operated by float-actuated switches. All power used in raw wastewater pumps will be metered through a separate KWH meter. The first cost of accomplishing raw wastewater pumping totaled $376,000, equal to $5.12 per P.E.

(4) Carrousel Activated Sludge - The external tank walls and dividing walls are of a simple cantilever design. The tank bottoms between walls are paved with five inches of mesh-reinforced concrete placed on top of a 30-mil pvc membrane. The membrane is required to assure watertight construction. Three of the six bridge-mounted aerators are 75 hp; three are 60 hp. All aerators are two speed allowing for power flexibility. The aerators are mounted to allow vertical adjustment of im-

peller submergence, and thirty foot long effluent weirs are also manually adjustable to provide further flexibility in controlling actual energy consumption. Power supply to all aerators is metered with a separate KWH meter.

The three Carrousel tanks are arranged with valves and piping to permit parallel or series operation, and any of the three tanks may be taken out of service while the others remain in operation. Also, a sampling bridge across all three tanks permits instantaneous measure of dissolved oxygen profiles with a portable DO meter. The three Carrousel tanks have a combined total volume of 7.2 million gallons, providing a retention time in excess of 40 hours, not including returning sludge volumes. The construction cost of the Carrousel activated sludge tanks, aerators, electrical work, pipe, valves, and etc. totaled $2,042,000, equal to a unit cost of $27.79 per P.E.

(5) Final Clarification - Figures 4, 5 and 6 show the layout of the three 80 foot diameter clarifiers. These provide a surface loading of 278 gallons/day/sq.ft., based upon design flow of 4.2 MGD. Accordingly, both the Carrousel tanks and the clarifiers can accommodate peak (wet weather) flows in excess of three times average flow without upsetting plant operation. The clarifiers use the simplest form of sludge collector. As shown in Figure 5, the clarifier bottom is paved with 5 inch membrane concrete after the walls and center piers are constructed. This design yielded a low cost for clarifier construction, and also eliminated the need for "grouting in" of the clarifier bottoms. The construction cost of three clarifier tanks including sludge collectors and related pipe, valves and etc. amounted to $444,000, equal to a unit cost of $6.04 per P.E.

(6) Activated Sludge Return - This important plant function consists merely of a method of providing pumps and piping to return sludge to the Carrousel activated sludge process. Submersible pumps are used to pump return activated sludge. Each activated sludge return pump chamber contains three submersible pumps, each rated at 40 percent volume of flow to the tank. This will allow sludge return at a rate of 40, 80 or 120 percent of plant flow. The actual rate of return can be further adjusted by throttling. Total energy used in return sludge pumping will also be measured continuously by a KWH meter. This is a simple arrangement for the function, yet provides considerable flexibility. The pumps also provide a sub-function, as they will be used to dewater the clarifiers if required. The submersible pumps can be easily removed for maintenance or replacement. The cost of facilities for returning activated sludge is $164,000, equal to $2.23 per P.E.

(7) Skimmings Removal - The three clarifiers are equipped with conventional skimming equipment. Skimmings drop into the skimmings pump station, and two submersible pumps are provided

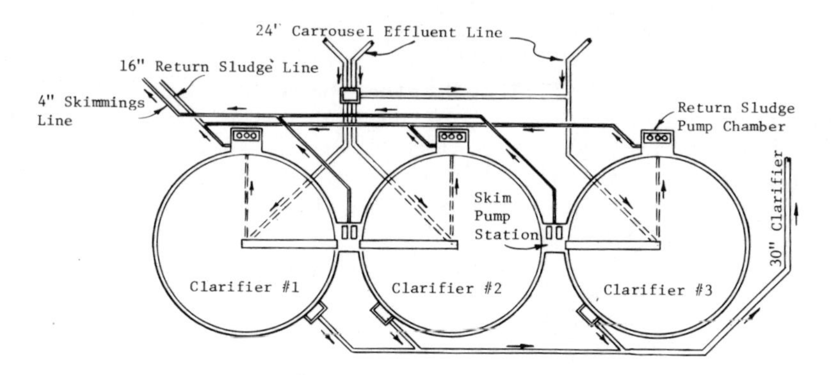

FIGURE 4.

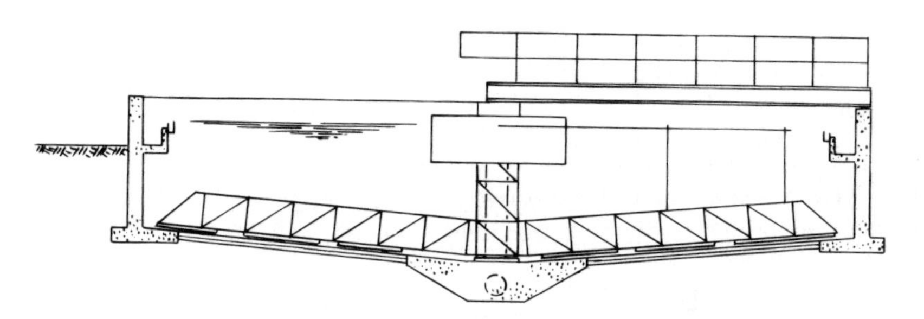

FIGURE 5. Clarifier Cross-Section

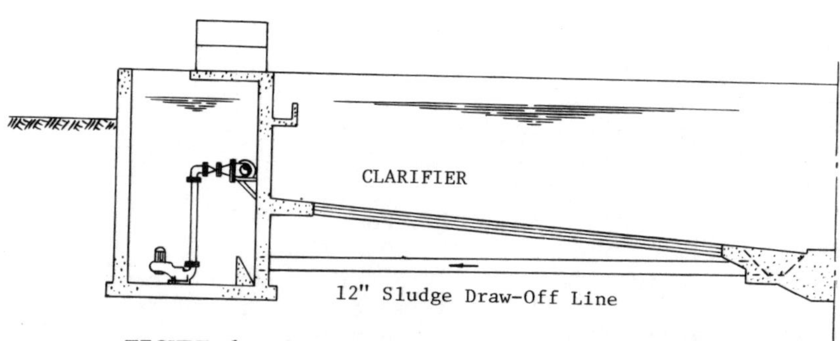

FIGURE 6. Return Sludge Pumping Chamber

to pump to the skimmings decant tank. Over the years we have observed this function to be one of the dirtiest operations in a treatment plant. The skimmings decant tank resembles a large open, baffled grease trap. It is fully fenced-in facility. In a few minutes each morning the operator will manually lift the skimmings from the surface of the decant tank and deposit same into a garbage can. At weekly intervals such skimmings will be hauled to the landfill. The construction cost of skimmings removal facilities is \$41,000, equal to \$0.56 per P.E.

(8) Effluent Chlorination - Figure 1 shows the location of the two chlorine contact tanks, constructed adjacent to the raw wastewater pumping station. A small concrete block building on top of the tank houses chlorination equipment which is automatically paced by the plant effluent meter. One-ton chlorine cylinders are handled with 3 hp hoist equipped with both vertical and horizontal drive. Each half of the chlorine contact tank may be taken out of service for periodic cleaning by draining the tank into the raw wastewater pumping station. The construction cost of effluent chlorination equipment and facilities is \$137,000, equal to \$1.86 per P.E.

(9) Waste Solids Disposal - Figure 1 shows the location of the waste solids flocculation, thickener, and thickened solids holding tanks. Gravity thickened solids (thickened with or without polyelectrolyte aids) will be hauled to farm lands. The waste solids process is based upon experience by others with Carrousel plants in Europe. The solids are highly stabilized and non-odorous. With the control of heavy metals at manufacturing plants, the hauling of stabilized solids for discriminate use as a fertilizer on farmlands is considered to be the most cost effective method of disposing of waste solids. We expect to be able to thicken the waste solids to 4 percent. The hauling truck has a capacity of 1600 gallons. In addition, two waste solids drying beds have been provided. These will be cleaned with a front end loader to provide removal. The combined total cost of facilities and equipment for waste solids disposal is \$265,000, equal to \$3.61 per P.E.

SUMMARY

Bids were received on November 22, 1977, with the low bidder being Brasfield and Gorrie of Birmingham, Alabama, in the amount of \$3,520,800. The bid proposal included a total of 57 items, most of which were lump sum prices. Table No. 1 shows construction costs prorated into the nine plant functions. In summary, the wastewater treatment plant has a cost of \$287 per pound of BOD, equal to \$47.92 per P.E. It has been customary for Carrousel plants to be evaluated on the cost per P.E. For those who want to compare plant costs on the gallon basis, it is necessary to use the "equivalent gallon" based upon 100 gallons per P.E. Therefore, the Campbellsville plant has a cost of 48 cents per equivalent gallon. We conclude that a value analysis approach should be a part of the initial design of a wastewater treatment plant, and that basic plant processes should be reviewed as "functions", with the consistent endeavor to accomplish the function at least cost, considering first cost, operating cost, and energy costs. Published data (not provided herewith) on Carrousel energy requirements already document its capability as an energy effective activated sludge process. The results at Campbellsville show that the Carrousel system can be constructed in the United States at low first cost.

TABLE I.

COST SUMMARY
CARROUSEL PLANT - CAMPBELLSVILLE, KY.
BID DATE - NOVEMBER 22, 1977

		COST			
	Plant Function	Total	Cost/ #BOD	P.E. Cost	%
1.	Activated Sludge	$2,042,000	$166.76	$27.79	58.00
2.	Final Clarification	444,000	36.26	6.04	12.61
3.	Wastewater Pumping	376,000	30.71	5.12	10.68
4.	Solids Disposal	265,000	21.64	3.61	7.53
5.	Solids Return	164,000	13.39	2.23	4.66
6.	Effluent Clarification	137,000	11.19	1.86	3.89
7.	Skimmings Removal	41,000	3.50	0.56	1.16
8.	Comminution	36,000	2.94	0.49	1.02
9.	Grit Removal	15,800	1.29	0.22	0.45
	TOTAL	$3,520,000	$287.68	$47.92	100.00

ACKNOWLEDGMENTS

We acknowledge the cooperation of the Mayor of the City of Campbellsville, Kentucky, and the members of the Campbellsville Water and Sewer Commission in authorizing that their General Manager, Mr. Marshall McGlocklin, visit Carrousel installations in Europe. This led to the City's acceptance of the Carrousel process as recommended by the engineer.

We also acknowledge the cooperation of the Commonwealth of Kentucky Department for Natural Resources and Environmental Protection and Region IV of the U.S. Environmental Protection Agency for allowing designs and specifications to deviate somewhat from so-called "standards". Without such intellectual understanding maximum cost effectiveness would not have been obtainable.

REFERENCES

1. John D. Wright - Notes retained from the CEC-AIA Value Analysis Workshop conducted February 24 - March 1, 1974 at Boston, Massachusetts.

13. NITROGEN CONTROL WITH THE ROTATING BIOLOGICAL CONTACTOR

Ronald L. Antonie. Manager, Technical Services, Autotrol Corporation, Milwaukee, Wisconsin

INTRODUCTION

The rotating biological contactor is a fixed film treatment system for removal of carbonaceous and nitrogenous matter from domestic and industrial wastewaters. It consists of corrugated plastic media which is mounted on a horizontal shaft and slowly rotated while approximately half immersed in wastewater. (See Figure 1).

Biological growth develops on the rotating surfaces and consumes the soluble organic pollutants in the wastewater. Rotation of the plastic media contacts the biological growth with the wastewater and exposes it directly to the air for absorption of oxygen, promoting aerobic biological treatment.

The rotating biological contactor (RBC) provides simple and stable operation in nitrogen control applications because of its freedom from interdependence with a final clarifier. Low power consumption and low mechanical maintenance requirements also contribute significantly to its overall cost effectiveness.

NITRIFICATION

The RBC process is currently in use on a number of nitrification applications for combined BOD removal and nitrification on primary effluents, and nitrification only of secondary effluents. Figure 2 shows an example of a process diagram for the 1.0 MGD RBC installation at Gladstone.

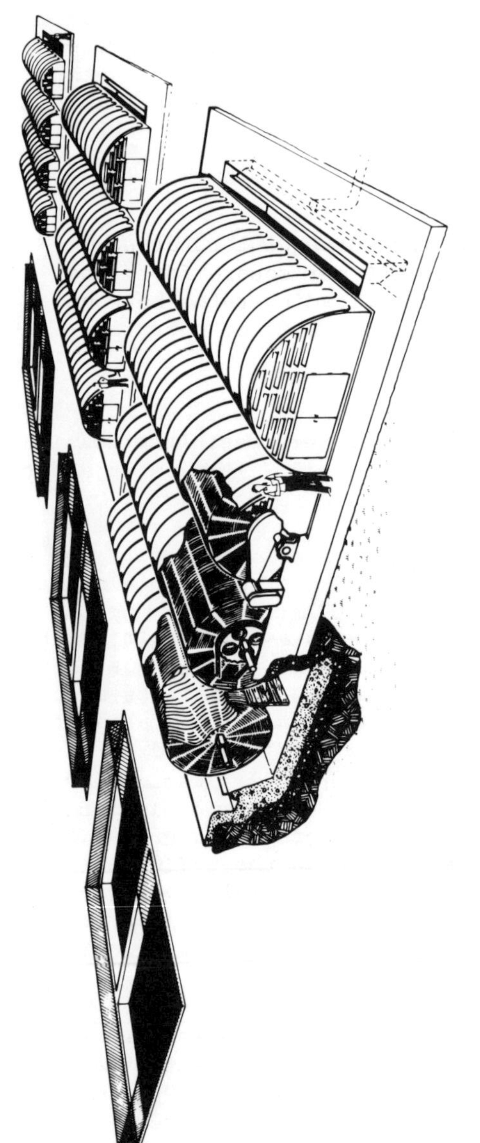

FIGURE 1. RBC TREATMENT PLANT

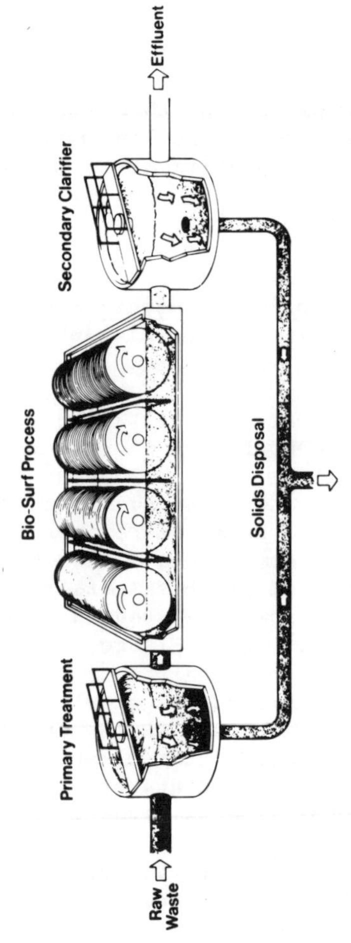

FIGURE 2. PROCESS FLOW DIAGRAM

Michigan. Although this installation was not designed to achieve nitrification, the design loading rate required for cold weather operation to achieve the required BOD removals results in an underloaded condition during the summer months and nitrification automatically occurs. At hydraulic loading rates above 3.0 GPD per square foot, the small amount of ammonia removal is due principally to synthesis of the soluble carbonaceous matter present. When the soluble BOD concentration is reduced to 15 mg/l, nitrifying bacteria can then grow rapidly enough to predominate on the subsequent rotating contactors. At loadings below 3.0 GPD per square foot, the rate of nitrification increases and at loading rates between 1.0 and 1.5 GPD per square foot, high degrees of nitrification were achieved.

Operation of the Gladstone plant during cold weather permitted the measurement of temperature effects on nitrification efficiency. This data is shown in Figure 3. Numbers adjacent to the individual data points equal the ammonia loading on the plant and the lines drawn through these data points are at approximately equal ammonia loading conditions. The effect of temperature can then be measured by moving up or down along a constant ammonia loading line. Above 60°F there does not appear to be a significant effect of temperature at the Gladstone plant. Below 60°F, the effect of lower temperatures can be readily seen and is very close to that measured on other types of fixed film nitrification reactors.[1]

To develop a more rigorous model for nitrification design, it was necessary to collect test data from individual stages from a multi-stage RBC plant which is nitrifying a domestic waste. Figure 4 shows test data from a two meter diameter pilot plant at Madison, Wisconsin. Samples were taken from each of the 4 stages of treatment in series and the specific ammonia removal rate was determined for each of the stages and plotted as a function of the ammonia nitrogen concentration in Figure 4. The data shows that ammonia removal by the RBC is first order with respect to ammonia nitrogen concentration up to approximately 5 mg/l. Above that concentration the removal rate appears to level out at 0.3-0.4 pounds per day per 1000 square feet. Above 5 mg/l staging will not be of any benefit because of the zero order removal rate. However, to produce effluents less than 5 mg/l, staging will be necessary to maximize the efficiency of the RBC surface area.

Similar test data was collected from a 0.5 meter diameter pilot plant at the City of Guelph, Ontario (Figure 5). At Guelph the test data shows the ammonia removal rate to increase with concentration, again, up to approximately 5 mg/l. Above

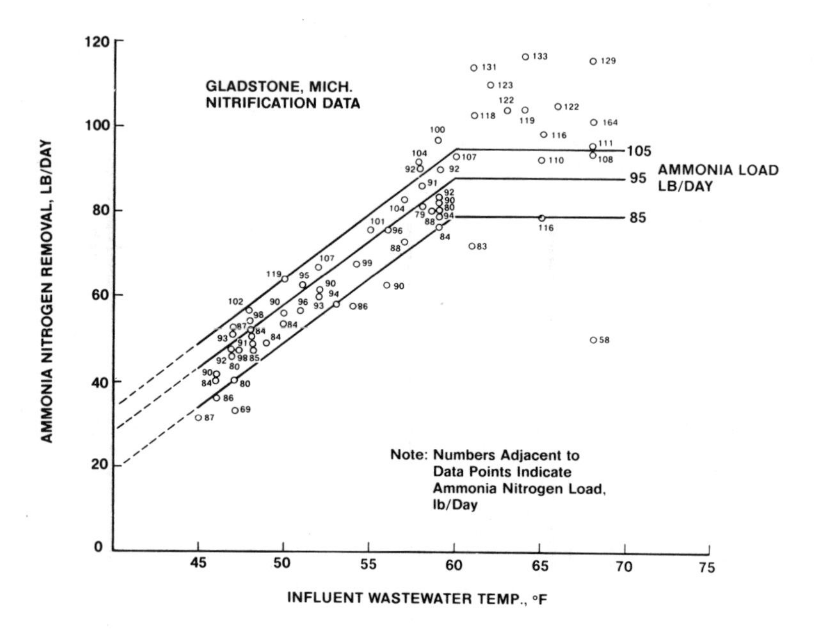

FIGURE 3. TEMPERATURE EFFECTS

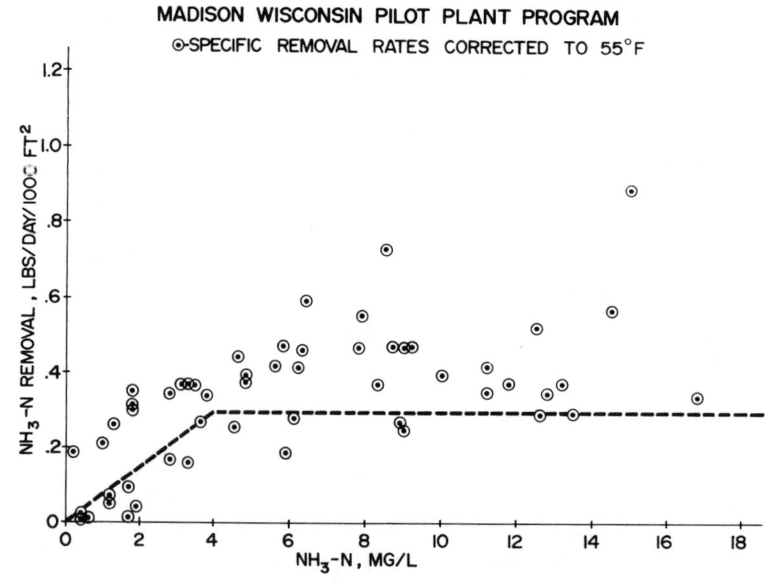

FIGURE 4. AMMONIA REMOVAL RATE/MADISON

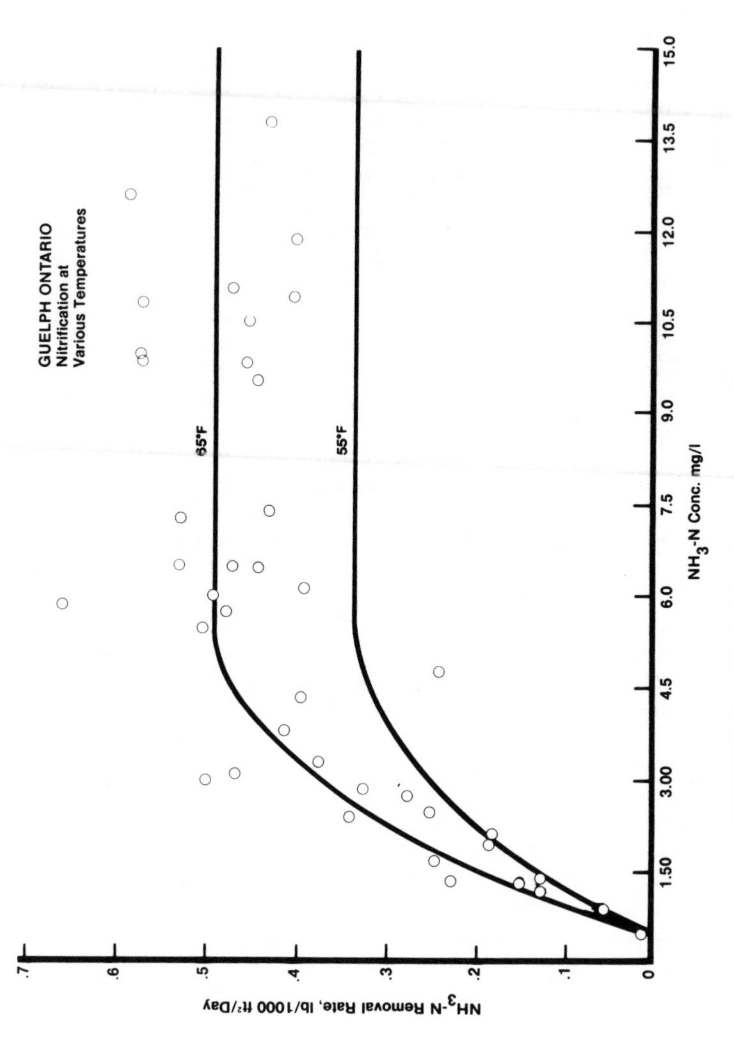

FIGURE 5. AMMONIA REMOVAL RATE/GUELPH

that concentration the removal rate remains constant. At Guelph the maximum removal rate was approximately 0.5 pounds per day per 1000 square feet at a temperature of 65°F. When this is adjusted to 55°F using the temperature effect from Figure 3, we find a relationship very similar to that measured at Madison, Wisconsin. That is, ammonia removal increases linearly with ammonia concentration up to 5 mg/l and a maximum rate of approximately 0.3 pounds per day per 1000 square feet and remains zero order above 5 mg/l.

The ammonia removal relationship determined from the Madison and Guelph pilot plants was then used to develop a design curve to cover a variety of nitrification requirements. The design technique utilizes the ammonia removal relationship shown in Figure 4 and 5 at 55°F and uses a multi-stage graphical technique described by Antonie[2] and McAliley[3]. Using the multi-stage design technique the final design curve shown in Figure 6 was developed. It shows effluent ammonia nitrogen concentration as a function of overall hydraulic loading on 4-stages of RBC treatment and parametric with influent ammonia nitrogen concentration for a wastewater temperature of 55°F. To use this design curve for a nitrification application on a primary treatment effluent, it will first be necessary to reduce the soluble BOD concentration to 15 mg/l using either the RBC process or some other secondary treatment method. For nitrification of a secondary effluent whose soluble BOD concentration is already less than 15 mg/l, the design curve of Figure 6 can be used directly.

Using the measured temperature effect in Figure 3 and a base temperature of 55°F the temperature correction factor in Figure 7 was developed. This correction factor should be used to reduce the design hydraulic loading determined from Figure 6 for those nitrification applications where low wastewater temperatures are expected. For applications where effluent ammonia standards are different for cold weather and warm weather operation, it will be necessary to design the process for each condition to determine which condition controls the overall design.

NITRIFICATION DESIGN CONSIDERATIONS

Because the RBC process employs a fixed-film of nitrifying bacteria, there is no need for sludge recycle. This significantly simplifies the operating of a nitrogen control plant when compared to a suspended growth type of system. There is no need to develop and control the long sludge age

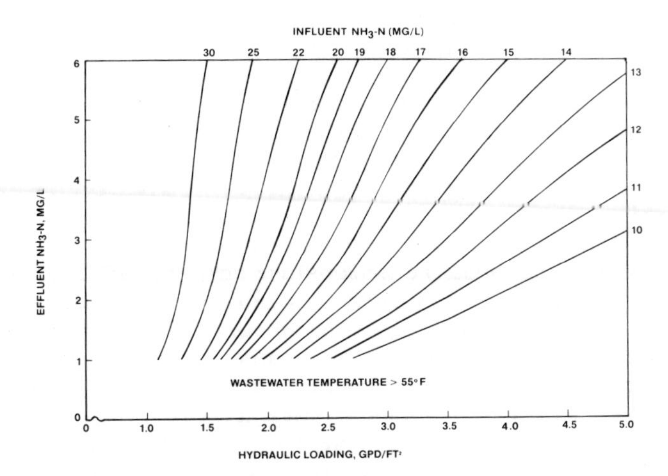

FIGURE 6. DESIGN CURVES/NITRIFICATION

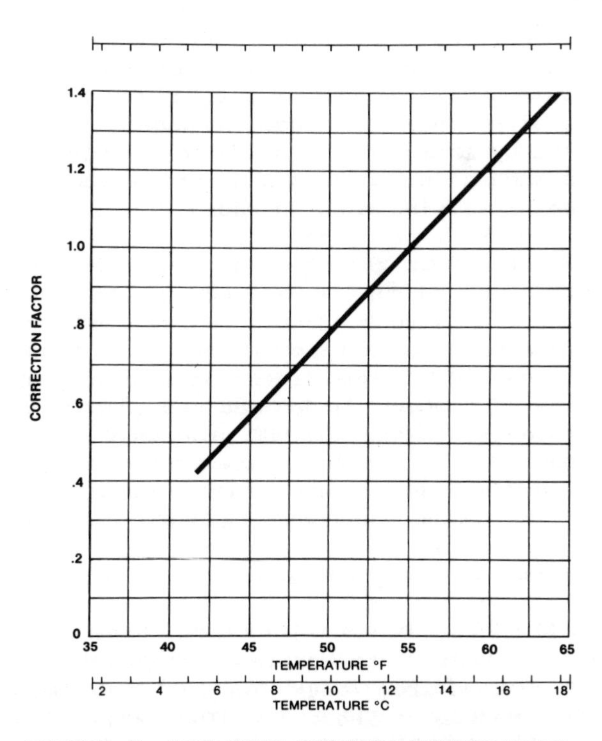

FIGURE 7. BIO-SURF PROCESS TEMPERATURE CORRECTION FOR NITRIFICATION

necessary for nitrification. It is not necessary to achieve efficient solids separation and thickening in the final clarifier and rapid recirculation back into the aeration tanks.

In addition to simpler operation, a fixed-film system also offers greater flexibility over a suspended growth system. From the design curve in Figure 6 it can be seen that it is possible to design the RBC system for essentially any desired effluent ammonia nitrogen concentration. In a suspended growth system, the necessity to design to a very long sludge age to achieve stable nitrification also results in a very high degree of nitrification. It is not practical to attempt to operate a suspended growth system for only a moderate degree of nitrification because of the relative instability resulting from the relatively short sludge age. Therefore, it is possible, when designing to various discharge standards for winter conditions, summer conditions or different discharge locations, to evaluate the relative cost effectiveness of the RBC process for these different conditions.

Many investigations have shown that when using the RBC process for nitrification, there is little or no net sludge yield from the fixed culture. Theoretically there would be approximately 3.5 mg/l of new nitrifying mass generated from each 20 mg/l of ammonia oxidized. However, in the RBC process, this is offset by endogenous respiration and by the activity of predators such as protozoans and rotifers which are consuming the nitrifying mass as rapidly as it can grow. Table 1 shows nitrification data from a full scale RBC pilot plant.[4] The data shows that the influent suspended solids and effluent suspended solids were essentially the same throughout the entire test program indicating little or no solids generation by the nitrifying culture. Since effluent solids will essentially equal influent solids, there is a significant benefit which can be accrued through elimination of clarification following the RBC process in a nitrification system. When nitrifying a secondary effluent where the influent suspended solids are within effluent, discharge standards, it will not be necessary to remove them from the RBC effluent before going directly to the receiving water. For applications nitrifying secondary effluents where the existing secondary effluents suspended solids level is not suitable for direct discharge, the RBC mixed liquor can go directly to tertiary filtration. Table 2 shows test results from filtration of RBC effluent from the same test program as Table 1.[5] The data indicates that at conventional filter loadings a high quality effluent can

TABLE I. OPERATING PERFORMANCE SUMMARY AIR ACTIVATED SLUDGE PILOT PLANT-NITRIFICATION MODE (SAS)

Period Dates Feed Unit		1 1/1-2/25 SOC	2 2/26-3/8 SOC	3 3/9-3/15 SOC	4 4/10-4/30 SRC	5 5/1-5/19 SRC	6 5/20-6/9 SRC	7 6/11-6/25 SRC	8 6/26-7/1 SRC	9 7/2-7/11 STF
Inf Flow	gpm	3.7	3.7	3.0	3.00	2.9	2.9	3.0	3.0	2.9
Recycle Ratio		0.51	0.73	0.63	1.27	1.69	1.68	1.68	1.64	1.59
Ret. Time (Q)	hr	3.11	3.24	3.90	3.86	3.94	3.91	3.90	3.86	3.98
BOD Inf Unf Inh	mg/l	14.3	9.6	8.3	43.6	41.0	32.5	44.6	55.3	81.7
BOD Eff Unf Inh	mg/l	5.3	2.9	2.6	5.4	16.3	10.8	7.5	5.8	8.4
BOD Rem Unf Inh	%	63	70	69	88	60	67	83	89	90
BOD Inf Filt Inh	mg/l	7.9	3.2	1.9	8.7	14.3	15.4	25.6	33.5	19.2
BOD Eff Filt Inh	mg/l	1.9	0.5	1.1	1.5	1.6	1.3	1.0	0.7	1.3
BOD Rem Filt Inh	%	76	04	42	83	89	92	96	98	93
NH_3-N Inf	mg/l	....	....	....	15.8	15.1	13.2	12.2	15.7	18.1
NH_3-N Eff	mg/l	....	....	....	0.1	0.3	0.1	0.0	0.0	0.0
NH_3-N Rem	%	....	....	....	99	98	99	99+	99+	99+
NO_3-N Inf	mg/l	0.30	0.18	0.73	0.35	0.54	0.23	0.40	0.28	0.10
NO_3-N Eff	mg/l	0.62	1.15	0.34	14.23	20.75	16.83	14.12	16.70	17.48
NO_2-N Inf	mg/l	0.29	0.23	0.46	0.17	0.31	0.14	0.16	0.13	0.09
NO_2-N Eff	mg/l	0.92	0.45	0.28	0.10	0.32	0.12	0.03	0.11	0.04
D.O.	mg/l	5.7	5.7	6.3	4.9	5.5	4.9	5.5	5.1	5.3
O.U.R.	mg/l/hr	9.2	7.3	4.9	31.7	29.8	22.2	19.3	26.1	26.1
MLSS	mg/l	1,355	2,580	560	7,180	4,085	3,430	2,655	2,800	2,425
MLVSS	mg/l	955	2,200	540	5,000	2,990	2,280	1,935	1,995	1,800
TSS Recycle	mg/l	3,900	5,515	2,010	12,730	6,390	5,235	4,025	4,455	4,095
TSS Inf	mg/l	28	21	23	153	172	119	86	87	136
TSS Eff	mg/l	21	38	26	25	76	54	30	76	81
Tot Sludge Prod	lb/da	-0.35	-0.98	0.49	-1.16	-2.63	-0.15	0.85	....	-1.80
Mixed Liquor Temp	°C	12.4	14.4	13.9	16.0	17.3	19.1	19.9	20.3	23.7
pH Inf		7.2	7.3	7.2	7.6	7.7	7.7	7.8	7.8	8.0
pH Eff		8.0	7.5	7.5	7.5	7.4	7.6	7.6	7.6	7.5
Alkalinity Inf	mg/l	315	299	296	356	363	362	314	354	350
Alkalinity Eff	mg/l	374	315	308	217	184	219	215	212	188
Solids Wasted	lb/da	0	0	0	4.38	1.92	2.03	2.32	0.79	2.50
Sludge Age	da	10.10	9.73	4.83	8.32	5.52	5.30	4.74	11.64	4.17
Clar O. F. Rate	gpd/sf	853	842	680	687	672	678	680	682	666
Clar Solid Load	lb/sf/da	10.5	21.2	3.6	63.9	43.1	35.9	28.0	29.3	24.3
ZSV	ft/hr	16.4	9.9	....	....	6.9	15.3	10.9	6.5	4.9
SVI	mg/l	75	71	93	76	44	43	75	123	153

TABLE II. GRAVITY FILTRATION OF ROTATING BIOLOGICAL CONTACTOR EFFLUENT

Dates 1974	No. of Runs	Media	Flow gpm/sf	Length of Runs Hr.	Head Loss Ft.	TSS Inf mg/l	TSS Eff mg/l	Tot BOD Inh* Inf mg/l	Tot BOD Inh* Eff mg/l	Eff Turbidity JTU	Ave. Solids Loading lb/sf	Ave. Solids Load to H=4 ft lb/sf
4/17–5/3	19	24" Coal 12" Sand 9" Gravel	2.46	9.30 hr	4.64 ft	21	1.9	18	4	2.0	0.62	0.53
5/3–5/7	6	14" Coal 9" Sand 9" Gravel	2.52	14.64 hr	6.17 ft	18	2.0	20	8	2.7	0.55	0.35
5/7–5/10	7	Same As Above	3.83	10.70 hr	6.04 ft	15	2.0	18	5	2.3	0.69	0.46
5/10–5/14	6	6" Low Dens. Coal 14" Coal 9" Gravel 9" Sand	4.00	14.22 hr	6.96 ft	18	1.7	15	2	1.7	0.84	0.50
5/14–5/22	4	Same As Above	2.47	40.96 hr	7.30 ft	29	2.8	21	5	2.0	0.86	0.47
5/23–5/28	11	Same As Above	4.72	12.19 hr	6.11 ft	25	1.5	18	4	1.7	1.42	0.93
6/4–6/6	3	Same As Above	3.77	20.58 hr	6.14 ft	16	1.4	19	5	2.0	0.72	0.47
6/6–6/10	3	Same As Above	2.52	22.41 hr	6.70 ft	18	1.4	21	5	2.0	0.55	0.33
6/10–6/12	7	Same As Above	5.89	8.74 hr	4.90 ft	26	2.6	25	6	2.6	1.82	1.48

*Inhibitor add to suppress nitrification during incubation

be produced without having a clarification step between the RBC process and the tertiary filter.

The full scale RBC plant at Cadillac, Michigan has been employing this technique for several years and achieving final effluent qualities equivalent to those shown in Table 2.

When designing any biological process for nitrification, it is important that all ammonia loadings be considered in the design calculations. In addition to the ammonia nitrogen concentration in the plant influent, consideration must also be given to the ammonia nitrogen concentration in any return sludge liquors, particularly those from anaerobic digestion and thermal sludge conditioning. These liquors will often contain several hundred mg/l of ammonia nitrogen and if recycled under peak loading conditions will often result in a breakthrough of ammonia nitrogen in the final effluent. To avoid this condition, it is necessary to control the return of sludge liquors so that they coincide with off-peak loading conditions. This will permit the biological process to absorb the ammonia loading and prevent any deterioration in effluent quality.

UPGRADING EXISTING SECONDARY TREATMENT PLANTS FOR NITRIFICATION

Because of its modular construction, low hydraulic head loss, and shallow excavation, it is possible to incorporate RBC process equipment into many existing treatment plants with little or no change in existing facilities.

TRICKLING FILTER PLANTS

Trickling filter plants can be upgraded in a variety of configurations using RBC equipment. RBC equipment can be placed in parallel or in series with an existing trickling filter. In parallel operation the loading to the trickling filter is reduced so that it can achieve both BOD removal and nitrification and the RBC process in parallel with it also achieves both BOD removal and nitrification. The two effluents are then combined and flow into the existing final clarifier if it is of sufficient capacity. A more common approach is to place the RBC equipment in series with the existing trickling filter. The trickling filter is then used for roughing treatment and the RBC process provides the balance of the BOD removal and the nitrification. RBC effluent flows directly to the existing final clarifier. If it is not practical to place the RBC equipment between the existing trickling filter and final clarifier, the final clarifier effluent can then be treated directly by the RBC process and because of the negligible solids production during nitrification, the effluent may often be able to be directly discharged to the receiving water. In some applications where the trickling filter capacity is small when compared to ultimate treatment requirements, it is more cost effective to convert it to another use such as a secondary clarifier, sludge thickener or storage tank.

UPGRADING EXISTING ACTIVATED SLUDGE PLANTS

RBC equipment can be placed in parallel or in series with existing aeration tanks. Parallel operation reduces the loading on the aeration system so that it can achieve both BOD removal and nitrification. The RBC equipment would be designed for both BOD removal and nitrification and discharge its effluent into the existing secondary clarifier. RBC equipment can be placed before existing aeration tanks to provide roughing treatment, reducing the BOD concentration to the point where a stable sludge age for nitrification can be developed in the subsequent aeration step. The most common application is to treat the activated sludge effluent where the RBC equipment would be providing nitrification with the effluent being either directly discharged to the receiving water or to a tertiary filtration step. A recent process development called the Surfact process has RBC units placed directly into existing aeration tanks to combine the treatment of the fixed RBC culture and the existing suspended growth system. Doing this extends the overall sludge age of the system and improves its settleability to the point where the sludge age can be extended to include both BOD removal and nitrification. One of the significant benefits of this application is that with air driven contactors, the existing air supply needs little or no expansion to achieve both BOD removal and nitrification.

DENITRIFICATION

When a rotating biological contactor is operated in a completely submerged position and fed with nitrified effluent and a carbon source, the surfaces will become covered with denitrifying bacteria. Operating in the submerged position prevents the normal aeration mechanism and permits the anaerobic denitrifying bacteria to predominate. Research work on this application has been done by Murhpy et al.[6] the Chicago Metropolitan Sanitary District[7], and Ishiguro[8]. The data developed by Murphy is shown in Figure 8 where the rate of denitrification per unit RBC area is shown as a

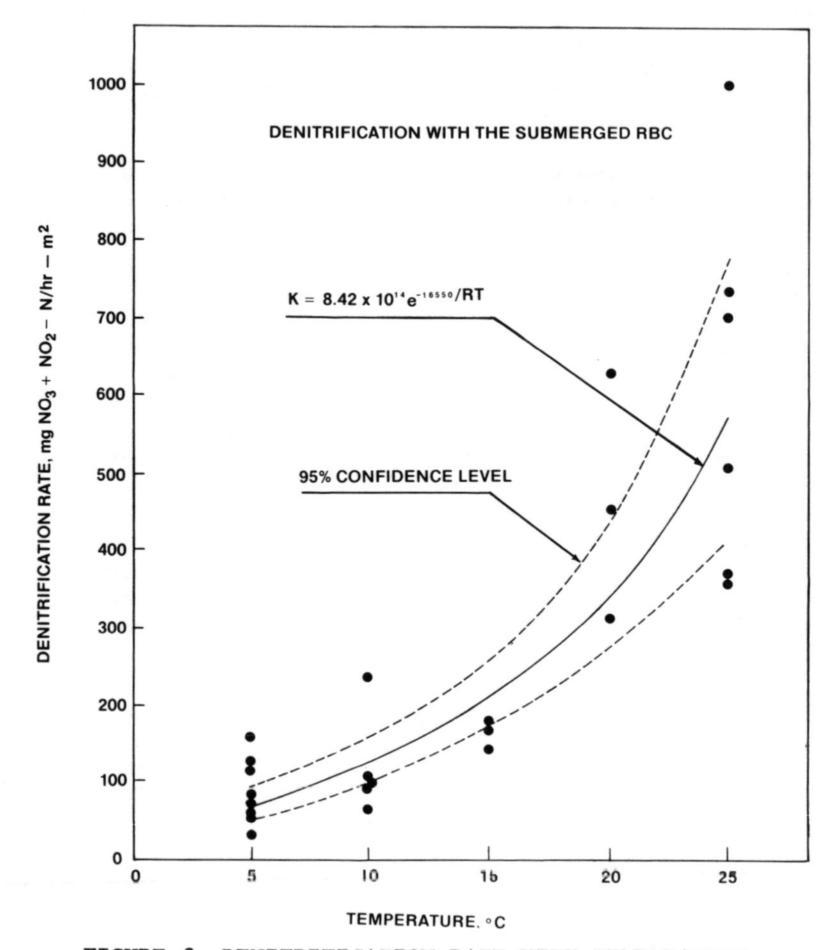

FIGURE 8. DENITRIFICATION RATE WITH TEMPERATURE

function of wastewater temperature. The denitrification rate for a temperature of above 12°C is approximately 1 pound per day per 1000 square feet of contactor area. This removal rate was used to construct the denitrification design curve in Figure 9. The design curve shows effluent nitrate nitrogen concentration as a function of hydraulic loading in gallons per day per square foot and is parametric with influent nitrate nitrogen concentration.

Using the measured effect of temperature in Figure 8, the temperature correction factor for design purposes was developed in Figure 10 for wastewater temperatures above or below 55°F. The hydraulic loading from Figure 9 should be increased or decreased accordingly.

The rotating biological contactor can then be used to provide all three steps of biological treatment, BOD removal, nitrification and denitrification. For a small treatment plant, such a configuration is shown in Figure 11. Here three contactors are arranged to provide all three steps of biological treatment. Two aerobic units operate in parallel to provide the BOD removal and nitrification. The nitrified effluent then flows directly to a third submerged unit with methanol addition to achieve the denitrification. The denitrified effluent containing all of the waste biological solids from all three steps of treatment then passes to a final clarifier.

For a large installation the configuration in Figure 12 would be employed. Here the wastewater flow is perpendicular to the shaft assemblies. After the wastewater flows through several aerobic units in series, achieving both BOD removal and nitrification, it then flows over a weir into completely submerged units for the denitrification. For both small and large plants, a floating plastic cover can be placed on the liquid to minimize the surface absorption of oxygen and to minimize the methanol consumption.

DESIGN CONSIDERATIONS FOR DENITRIFICATION

Because of the low sludge yield in the RBC process, treatment plants requiring all three steps of treatment need not have clarifiers between steps. Mixed liquor solids from BOD removal can pass through nitrification and denitrification without affecting either reaction. Sludge yield on nitrification is negligible and sludge yield on denitrification is quite small so that the final effluent from all three steps of treatment will typically contain

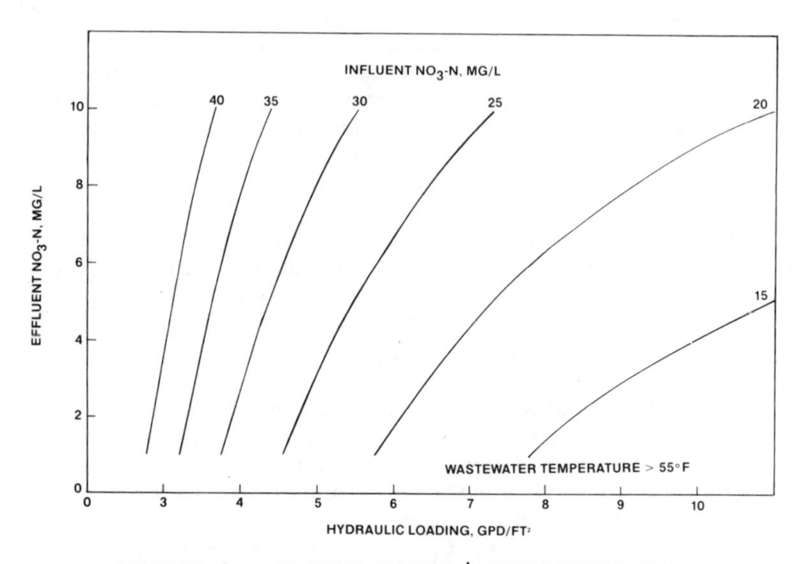

FIGURE 9. DESIGN CURVES/DENITRIFICATION

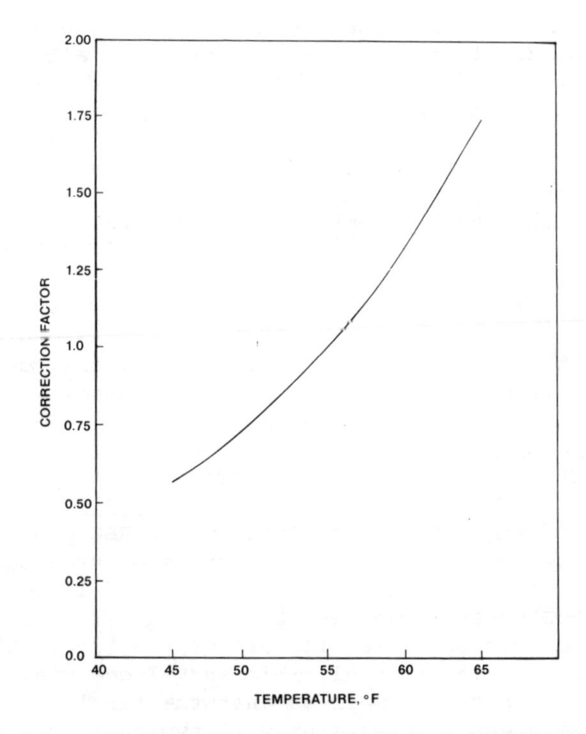

FIGURE 10. TEMPERATURE CORRECTION FOR DENITRIFICATION

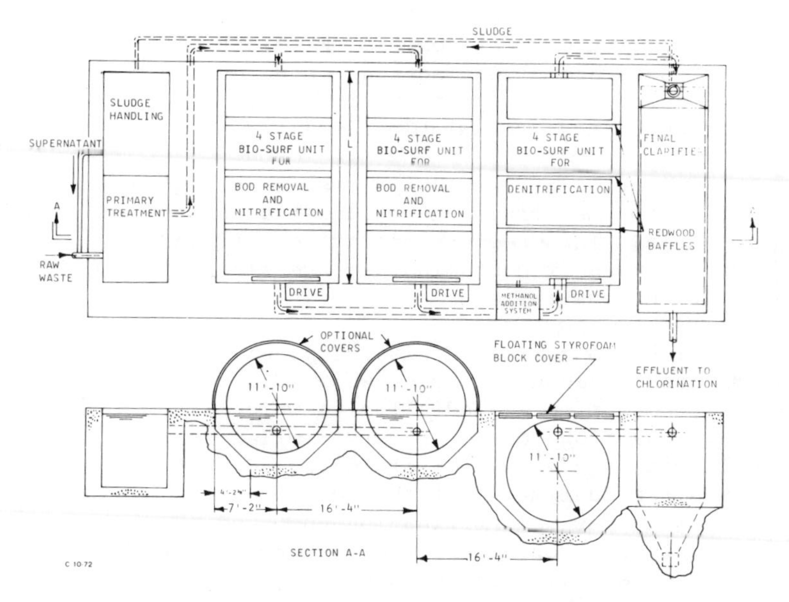

FIGURE 11. BIO-SURF PROCESS/SMALL SYSTEM

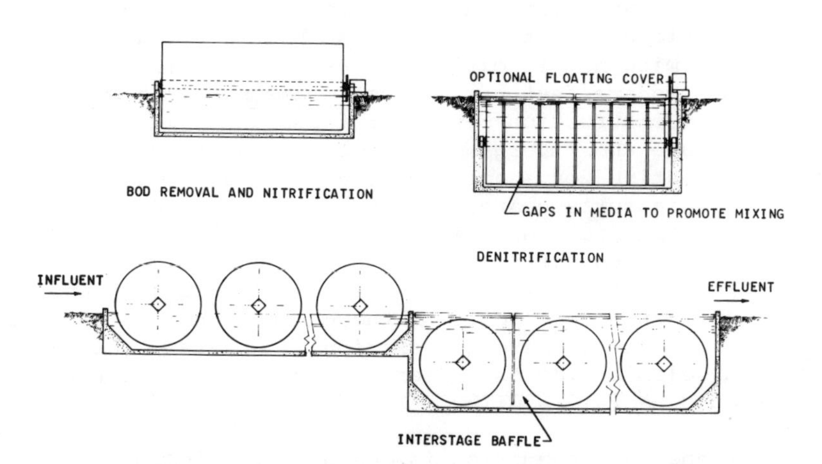

FIGURE 12. BIO-SURF PROCESS/LARGE SYSTEM

less than 150 mg/l of solids. A single clarification step at the end of the plant which is designed just as it would if it were providing clarification after a BOD removal step, significantly reduces the cost of the clarification function for the overall treatment scheme. Elimination of the need to control separate sludge ages as is necessary for suspended growth types of systems greatly simplifies plant operation and can result in a more stable final effluent.

The operation of the RBC process as a staged system for denitrification maximizes the utilization of the methanol carbon source. This reduces operating cost and minimizes the possibilities of contaminating the final effluent with excess carbanecous matter.

The first commerical scale RBC plant for denitrification is at Miyazaki, Japan and is shown in Figure 13. It is treating runoff from a landfill site which is rich in ammonia nitrogen. The flow scheme is very similar to that shown in Figure 11 with the exception of adding a final RBC unit to polish the residual methanol through aerobic biological treatment following the denitrification unit and preceding the final clarifier. Monitoring the performance of this unit confirms the denitrification rate of 1 pound per day per 1000 square feet when producing effluents of less than 5 mg/l nitrate nitrogen.

CAPITAL AND OPERATING COSTS

RBC systems designed to nitrify secondary effluents and the nitrification portion of an RBC system in a combined BOD removal and nitrification system would utilize high density media. This media configuration has closely spaced surfaces made possible by the thin biological cultures that develop when performing nitrification. These units will contain up to 150,000 square feet per 12 foot diameter by 25 foot long assembly. Construction costs for these units including concrete tankage and fiberglass enclosure and including freight and installation costs amounts to approximately 35¢ per square foot of media surface area. Standard media assemblies with 100,000 square feet per unit which would be used for BOD removal preceding nitrification and would also be used in a submerged position for denitrification would cost approximately 45¢ per square foot. Although the concrete tankage required for the submerged operation is greater, this is offset by eliminating the fiberglass cover and the total cost should be very similar.

Low energy consumption is one of the principal benefits of the RBC process when compared to alternative systems. For nitrification applications the energy consumed is about 2.5 HP per 100,000 square feet. For combined BOD removal and nitrification applications, it is approximately 3.0 HP per 100,000 square feet. This is approximately 1/4 to 1/2 that of the competitive biological systems and when combined with construction costs very often leads to a lower total cost of treatment. Low construction and operating cost and its simplicity of operation have gradually increased acceptance of the RBC process for nitrogen control applications throughout the U.S. during recent years. There are now many nitrogen control applications employing this process, covering a wide range of treatment plant sizes. The largest installation for combined BOD removal and nitrogen control is at Columbus, Indiana with a capacity of 12.5 MGD. It is shown in Figure 14. It contains 80 individual shaft assemblies and is the largest operating RBC plant in the world.

Figure 15 shows the RBC installation at Hinckley, Ohio for nitrification of activated sludge effluent. This plant has a capacity of 2 MGD. The largest RBC application for nitrification of activated sludge effluent is under construction at Peoria, Illinois and contains 84 shaft assemblies which will be treating 37 MGD of activated sludge effluent.

FIGURE 13. MIYAZAKI JAPAN

FIGURE 14. COLUMBUS, OHIO

FIGURE 15. HINCHLEY, OHIO

REFERENCES

1. Process Design Manual for Nitrogen Control. U.S. Environmental Protection Agency, Office of Technology Transfer (October, 1975).

2. Antonie, Ronald L. "Rotating Biological Contactor for Secondary Wastewater Treatment." Presented at Culp/Wesner/Culp Wastewater Treatment Seminar, So. Lake Tahoe, State Line, Nevada (October 27-28, 1976).

3. McAliley, Jr. Eugene. "A Pilot Plant Study of a Rotating Biological Surface for Secondary Treatment of Unbleached Kraft Mill Waste." Tappi, 57: No. 9, 106 (September, 1974).

4. "Advanced Wastewater Pilot Plant Treatment Studies." Conducted for the Consolidated City of Indianapolis by Reid, Quebe, Allison, Wilcox & Associates, Inc. (January, 1975).

5. Antonie, Ronald L. Fixed Biological Surfaces - Wastewater Treatment Cleveland, Ohio: CRC Press (1969).

6. Murphy, K.L. et al. "Nitrogen Control: Design Considerations for Supported Growth Systems." Journal Water Pollution Control Federation, 49: No. 4, 553 (April, 1977).

7. Prakasam, T.B.S., Robinson, W.E. and Lue-Hing, C. "Nitrogen Removal from Digested Sludge Supernatant Liquor Using Attached and Suspended Growth Systems." Proceedings, 32nd Purdue Industrial Waste Conference W. Lafayette, Indiana, 745 (May 10-12, 1977).

8. Ishiguro, Masayoshi. "Advanced Treatment of Sewage By R.B.C." Journal of Water and Wastewater (Japanese), 19: No. 7 805 (1977).

INDEX